Guide To Electric Power Transmission

Guide To
Electric Power
Transmission

A. J. Pansini
K. D. Smalling

Copyright © 1998 by
PennWell Publishing Company
1421 S. Sheridan Road/P.O. Box 1260
Tulsa, OK 74101

Library of Congress [CIP data]

Printed in the United States of America.

1 2 3 4 5 02 01 00 99 98

Contents

Preface

This book comprises the second revision of an original work entitled, *Basics of Electrical Power Transmission*. The revision includes updating of some material, expansion of the direct-current transmission discussion, and a new chapter on real-time capability.

"Basic Electricity" (formerly chapter 6) is moved to the appendices and replaces the previous appendix C, "U.S.-Metric Relationships." Appendix D, "Electrical Power Glossary," has been added to further enhance the use of this book by those inexperienced in the electric utility industry.

Basic information is provided for those just entering this field. Others with more experience may discover some valuable nuggets, as well as finding a desirable review. This text is also meant to serve the needs of those whose work is associated with this part of the electric utility industry. It is a training resource for the several categories of employees in electric utilities, both private and public.

This book is also a source of much-needed information for people in associated manufacturing and service industries. It should prove useful in schools and educational centers and to planning and civic groups interested in the aesthetic and economic development of reliable electric supply. The information presented here will benefit the financial community of investors and promoters, members of the legal profession involved in certain contracts and litigation, and arbitrators and mediators.

K. D. Smalling
Northport, NY

A.J. Pansini
Waco, Texas

Acknowledgments

Acknowledgment is made to the many manufacturers and agencies who contributed drawings, photographs, and descriptive material for this book. Acknowledgment is made as well to the many who have enabled these writers to assemble the information passed on in this work, as well as the staff of PennWell for their editing and artwork assistance in the preparation of the original manuscript. And last, but certainly not least, to our wives for their patience, understanding, and encouragement in the preparation of this work.

1

General Concepts

Like air, water, and food, electrical energy has become an integral part of our daily personal and business lives. It is so taken for granted that little thought is given to the process that produces and delivers this energy to where it is used. It is a unique process among industries; practically all that is produced is used instantly in the quantities required for light, actions, or communication, and is not stored in warehouses or on shelves.

Similar to other businesses, means must be provided to bring the product to the consumer—in this case, from the generating plant to the consumer's premises. In most business processes, provision is made for the delivery of the product in bulk and for its retail distribution. In an electrical power system, the bulk delivery is referred to as the *transmission system*, while the retail delivery is known as the *distribution system* (see Fig. 1-1).

Some people are unaware that an electric utility is a business enterprise, whether or not the industry has been deregulated in their state. Where the industry is regulated, the price charged for its product is not entirely under its control, but is fixed at various levels by regulatory bodies. This often leads to a public perception that an electric utility monopoly is guaranteed a profit no matter what. In fact, it is not guaranteed a profit. A limit is set on how much profit can be made under proficient management. The fact remains that utilities must pay for materials, labor, and capital they require and pay taxes just like any other business.

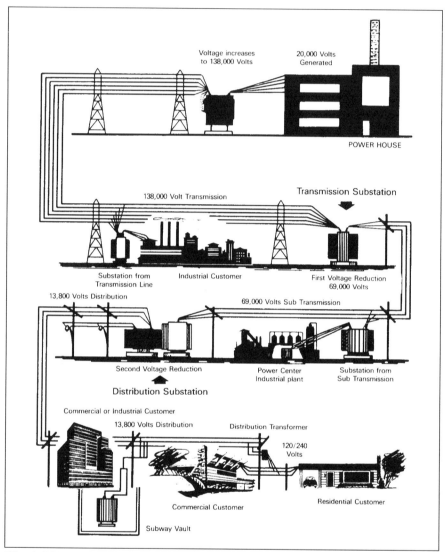

Fig. 1-1 The electrical supply system: electrical service from the generator to the customer

In obtaining these commodities, utilities must compete for them at prices generally dictated by the marketplace, while the prices charged for their electrical energy product are limited by regulatory agencies. In addition, electrical utilities must abide by regulatory standards for reliability of delivery and for the quality of the voltage at which the energy is delivered.

The Transmission System

Transmission systems carry the bulk supply to load centers from generating plants. These plants are generally located in outlying areas because of their environmental impact and need for large amounts of water for cooling purposes. A factory or generating plant is rarely situated at or near a population or load center. Generally, high voltages and great capacity are essential for the transmission system.

Transmission lines also provide interconnections for the transfer of electrical power between two or more utilities for economic and emergency purposes. Such interconnections of utilities are referred to as *pools or grids*, and sometimes more formally as *integrated systems*.

From wholesale centers or transmission substations, means are provided to distribute the electricity to regional load and population centers where a demand for the product exists. These means constitute the distribution system.

The principal purpose of a transmission system, then, is to carry bulk quantities of electrical energy to or between convenient points. At these points, the electrical energy may be subdivided for eventual delivery to one or more distribution systems. A synthesis of definitions accepted by the Federal Energy Regulatory Commission and various state utility commissions follows:

> A transmission system includes all land, conversion structures, and equipment at a primary source of supply; lines, switching and conversion stations between a generating or receiving point and the entrance to a distribution center or wholesale point; and all lines and equipment whose primary purpose is to augment, integrate or tie together sources of power supply.

The FERC also emphasizes the significance of transmission systems in a survey:

> The strategic importance of transmission is much greater than is indicated by the 20% average share in the overall cost of electricity. (The 20% average would pertain to transmission no matter who "owns" it.) Adequate interconnections, where

economically justified, provide the keys to large-scale generating units, to major savings in capacity because of load diversity, and the most efficient utilization of existing generating capacity. In short, interconnection is the coordinating medium that makes possible the most efficient use of facilities in any areas or region.

In Figure 1-1, note that the generator produces electricity at a pressure of 20,000 volts. This pressure is raised, by means of an apparatus known as a transformer, to a value of 138,000 volts or higher for the long transmission journey. This electric power is conducted over 138,000-volt (138-kV) transmission lines to substations located in important centers of population or electrical loads in the territory served.

When the electric power reaches the substations, its pressure is reduced or stepped down (also by a transformer) to 69 kV. It is then transmitted in smaller quantities to other substations in the local load areas. (In some cases, it might be stepped down to 13.8 kV for direct distribution to local areas.) Transmission circuits of such voltages may consist of open wires on poles in outlying zones (along highways, for example) where this type of construction is practical. Or it may consist instead of cables installed in ducts underground or buried directly in the ground in more densely populated areas.

While the transmission line system shown in Figure 1-1 is simple, many transmission line installations can provide for an interchange of power between two or more neighboring utility companies to their mutual advantage. Such lines permit one utility to purchase power from another when it is economically advantageous to do so. A utility might also purchase power produced elsewhere during periods of emergency when its own generating units may be out of service for repair or maintenance. Such interconnecting grids between utility companies may be very complex (see Fig. 1-2).

Wheeling of Power

The traditional concept of transmission interconnections between adjacent utility systems to exchange emergency power, economic generation, and capacity sales has changed under deregulation of the power industry. These transmission systems may now transfer energy and power requirements between any supplier and a consumer. Suppliers may be electric utility systems (sometimes geographically remote from the ultimate customer),

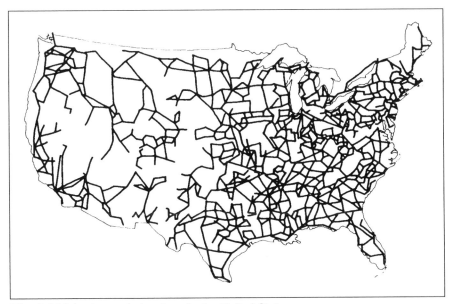

Fig. 1-2 Major transmission lines in the United States

independent power producers, or marketing firms (similar to stockbrokers).

The basic concept remains the same: using transmission to wheel power from the supply to the consumer with appropriate compensation for the use of the lines. However, the marketing aspect has been influenced by competition, open access to transmission, deregulation, and restructuring of utility corporations.

In the United States, deregulation was initiated through the federal Energy and Policy Act of 1992 (EPACT), introducing competition to the interstate and intrastate electric energy markets. In April 1996, the FERC issued Order 888, which eased the restrictions on transmission utilities. This was accomplished by opening access to transmission lines and separating the functions of generation and transmission for wholesale wheeling. Further development of competition in the electric energy market is now embodied in specific programs established by state regulatory authorities.

The wheeling of power under both normal and emergency conditions must still be planned and operated in such a manner as to provide reliable electric service. This requires adhering to generating reserve criteria and the operating capacity of the transmission system, particularly when the emergency operating capability limits are used. Notice of contingencies can range from several minutes or more (such as an operating generator unit coming off the line for some reason) to fractions of a second. The latter might occur when a transmission line trips under a short circuit condition—either temporarily or until repairs can be completed.

Wheeling of power on an interconnected system, with the emergence of competition in the supply of electric energy to consumers, has made planning and operating transmission systems a very complex procedure. There are more unknowns relating to quantity and timing than ever before. System operators must now cope with many more transactions on their system and have to coordinate with new entities called regional independent system operators (ISOs). Another complicating factor is the growing number of independent power producers (IPPs) and energy market brokers developing and managing contracts between supplier and consumer.

Line Characteristics

Before discussing the physical and electrical characteristics of transmission lines, it may be well to consider an analogy between electric transmission systems and water systems.

Water/Current Analogy

The flow of electricity can be compared to the flow of water. Whereas water flows in a pipe, electric current is made to flow through conductors or wires (see Fig. 1-3). To move a definite amount of water from point to point in a given amount of time, a large-diameter pipe and low pressure may be employed to force it through. Alternately, a small-diameter pipe could be used with high pressure applied to the water. If the higher pressure is used, the pipe must have thicker walls to withstand the pressure.

The same rules apply to the transmission of electric current. In this case, the diameter of the pipe corresponds to the diameter of the wire, and the thickness of the pipe walls corresponds to the thickness of the insulation around the wire. (For more detailed information, refer to appendix C, "Basic Electricity.")

Voltage and Load-Carrying Capability

Transmission lines are designed to transport large amounts of electricity—usually expressed in ratings by kVA (1,000 volt-amperes) or MVA (1 million volt-amperes)—over relatively long distances. Transmission systems

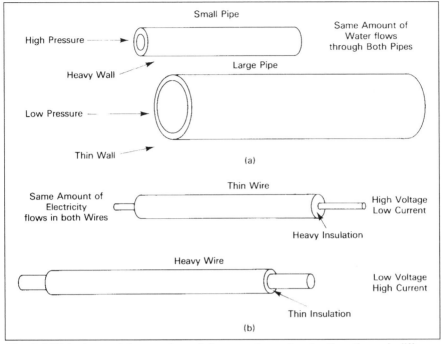

Fig. 1-3 Water/current flow analogy: **(a)** comparison of water flow through different size pipes **(b)** comparison of current flow in different size wires

are generally designed so that remaining lines pick up the load when a transmission circuit is interrupted. Normal line ratings are used under normal operating conditions. Emergency line ratings exceed normal ratings and become the design limit on the basis of known contingency conditions resulting from system load flow studies.

Transmission lines may carry less than 10 MVA to hundreds of MVA up to many hundreds of miles. When the *power factor* is known (refer to appendix C), the lines can also be rated in kW (kilowatts) or MW (megawatts) of electric power. The electrical pressures, or voltages, at which these lines operate may range from a few thousand volts to values of more than 765 kV, with experimental lines operating up to 1,500 kV. In general, the larger the amount of power to be carried, and the greater the distance to be traversed, the higher the voltage at which the transmission line is designed to operate (see Fig. 1-4).

Historically, standard transmission voltages have been approximate multiples of a base of 115 volts (see Table 1–1). Transmission lines strung

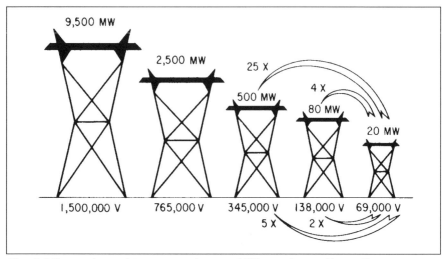

Fig. 1-4 Comparison of power-carrying capabilities and operating voltages

Table 1-1 Standard Transmission Voltages

13,800	23,000	34,500	46,000	69,000	115,000	138,000
230,000	354,000	500,000	765,000	1,000,000	1,500,000	

between substations and local or regional center are sometimes called *sub-transmission lines*. They generally operate at 69 kV or less. Lines operating at 500 kV or more are generally referred to as extra-high-voltage (EHV) lines.

Cost Considerations

Transmission facilities are costly, whether overhead or underground. The investment required, as a percentage of total utility plant, may be as much as 20%. (Again, that 20% is an average.) This is not only because of the requirement to move ever-larger amounts of power over considerable distances, but also because of the greater attention being given to appearance and effect on environment.

Transmission structures have been specifically designed for appearance, employing gracefully shaped and colored structures pleasing to the eye and blending with the surroundings (see Fig. 1-5). The color of insulators has generally been brown, but changes to light blue and gray standards have been made. These colors are less obtrusive against sky backgrounds and are aesthetically more acceptable.

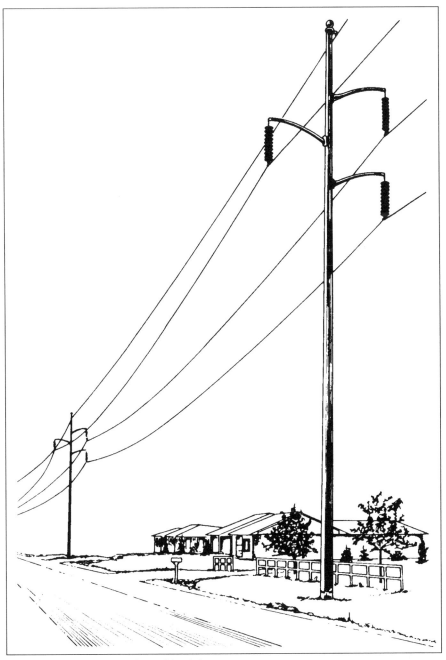

Fig. 1-5 Pole structures in residential areas

Transmission costs are affected by the costs of rights-of-way and the higher quality of construction needed in congested areas. Transmissions costs are also affected by the need for stronger ties both within and between utilities because of reliability requirements introduced by the use of larger generating units and underground cable systems. Transmission requirements may be reduced, however, when generating plants are in or near the load centers.

Data collected by the Federal Power Commission indicate costs for underground transmission lines in rural and suburban areas may vary from 10 to 30 times as much as overhead lines. Costs vary depending on the types of construction (duct or direct burial) and cables (solid insulation or hollow) employed. In urban areas, the ratio may be greater. In other places, because of rough ground, inaccessible terrain, and other special situations, this cost may be as much as 50 times greater (see Fig. 1-6). It would appear, therefore, that overhead lines may continue to predominate in transmission systems for some time to come.

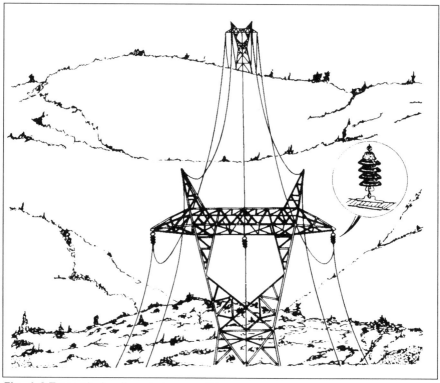

Fig. 1-6 Transmission lines in inaccessible areas

Overhead Vs. Underground Repairs

While overhead lines are exposed to the hazards of vandalism and weather, particularly lightning, ice, and wind, faults are usually relatively easy to find and repair. The reverse is generally true of underground lines. While relatively free from harmful exposure, when faults do occur, they are relatively hard (and time-consuming) to find and repair. Highly specialized skills are required for both their installation and repair. The addition of new lines or substations in an underground transmission system presents difficulties not usually met in overhead systems.

Rights-of-Way

Overhead transmission lines are admirably suited to open country areas where rights-of-way of ample width and reasonably straight lines are available (see Fig. 1-7). The straight-line route, with few angles, decreases the length of line required and the amount of guying or special construction necessary. Further, these rights-of-way are usually easier to acquire, whether by purchase, lease, or other arrangement. Appearance may not always be of paramount importance. Transmission line heights required by national and local safety code clearances for the voltage ranges involved and the economics of long spans make such overhead installations very adaptable to open, cross-country areas.

Rights-of-Way and Appearance

If transmission lines must be built in densely populated urban and suburban areas, rights-of-way may be restricted to relatively narrow areas along the streets or in alleys in the rear of buildings. Construction is usually limited to single (rather high) poles with span lengths of one city block or less. Unusual attention must be given to appearance, tree conditions, guying, and other factors that may be detrimental to good public relations. This may also include limited access in which materials may have to be transported by hand as well as agreements limiting work to off-schedule hours.

Tree Clearing

Often the rights-of-way require clearing of trees, brush, and other growth, and must be maintained sufficiently clear so that subsequent growth will not affect operation of the lines. The engineer will designate all danger trees, which may be removed or topped at the option of the contractor. In

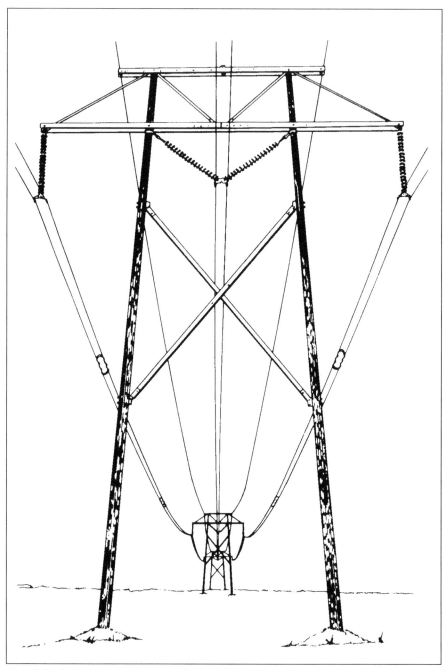

Fig. 1-7 Line installation in open, flat terrain

approximately level terrain, trees that would reach within five feet of a point underneath the outside conductor in falling are examples of danger timber. In wooded areas, the right-of-way is cleared back far enough so that trees falling will not fall into the line. Trees may be topped gradually so that so-called danger timber will not inflict damage to the lines (see Fig. 1-8).

Access roads must also be provided not only for construction, but for future patrols, inspections, repairs, and maintenance of the lines. Portions of the right-of-way must be cut so stumps will not prevent the passage of tractors and trucks along the right-of-way.

Rough Terrain

In some instances, such as where lines traverse heavily forested, rough terrain, helicopters may be used to deliver materials. Such deliveries often include preassembled tower structures, as well transporting personnel to the job sites for both construction and maintenance purposes (see Fig. 1-9).

Railroad Rights-of-Way

The installation of transmission lines along railroad rights-of-way would appear to provide a logical and desirable location for such lines (see Fig. 1-10). Often, however, these rights-of-way may also contain communication circuits that may suffer electrical interference from the high voltages of the transmission lines. Care must be exercised during work on such lines

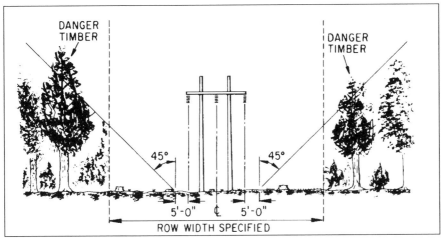

Fig. 1-8 Tree removal and topping along right-of-way

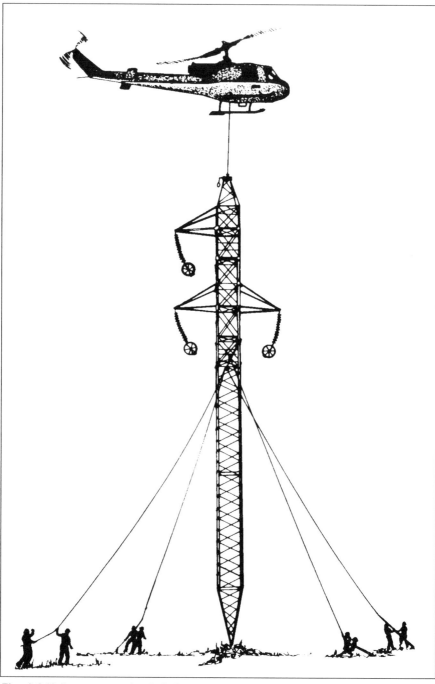

Fig. 1-9 Helicopters used to help install towers in inaccessible areas

Fig. 1-10 Transmission line (left) and low-voltage communications line (right) along railroad right-of-way

to insure the safety of personnel from passing trains. This may sometimes cause protracted and expensive delays in work schedules. Despite these drawbacks, railroad rights-of-way provide excellent routes for transmission lines. This is especially true since the railroad usually serves population or industrial centers that also constitute electric load centers.

Increasing Capacity

Overhead transmission lines more readily lend themselves to increased capacity because of system load growth than underground lines. Their capability may be increased by raising the operating voltage or by adding or replacing conductors. With the difficulty in securing construction permits because of environmental constraints and occasionally adverse public opinion, uprating existing lines is often much easier than building new lines.

Capacity additions to underground transmission lines are generally limited by initial design and installation. New techniques have improved the ability to uprate. These techniques include circulating insulating fluids in pipe-type cables or replacing the cable with larger conductors using new, thinner insulations at the same voltage level. Replacing the cable with a higher level of insulation to operate at a higher voltage is also possible with the new materials available as a result of industry research.

Another means of increasing capability is to apply real-time ratings to both overhead and underground lines. This is described in more detail in chapter 6. Such technology makes use of actual known data rather than assumptions in calculating the line rating.

Determining Transmission Voltages

Earlier, the relationship between conductor sizes and voltages in the transmission of electrical power was explained, including an analogy with water-carrying systems. This principle is basic in considering the choice of a voltage (or pressure) for a transmission system.

There are two general ways of transmitting electricity: overhead and underground. Both can use a variety of conductors. These include copper or aluminum conductors, aluminum conductors steel reinforced (ACSR), and, in some cases, steel conductors. But the insulation in the first instance is usually air except at the supports (towers, poles, or other structures), where it may be porcelain, glass, or other material. In underground transmission, the conductor is usually insulated with oil-impregnated paper, or a special type of plastic material.

In overhead construction, the cost of the copper or aluminum conductors as compared to the insulation is relatively high. It is therefore desirable, when transmitting large amounts of electric power, to resort to the higher electrical pressures or voltages, thereby employing thinner, less-expensive conductors that are easier to handle. Low voltages require heavy conductors that are costly, bulky, and expensive to install.

There is a limit, however, to how high the voltage and how thin the conductors can be. In overhead construction there is the problem of supports—poles, structures, and towers. If the conductor is too thin, it will not be able to support itself mechanically, and the cost of additional supports and insulators becomes inordinately high.

Underground construction faces the same economic limitations, and in this case, the expense of insulation. An underground cable must be thoroughly insulated and sheathed from corrosion. The greater the overall size of the cable, the more sheathing becomes necessary and more difficulty is experienced in its handling (see Fig. 1-11).

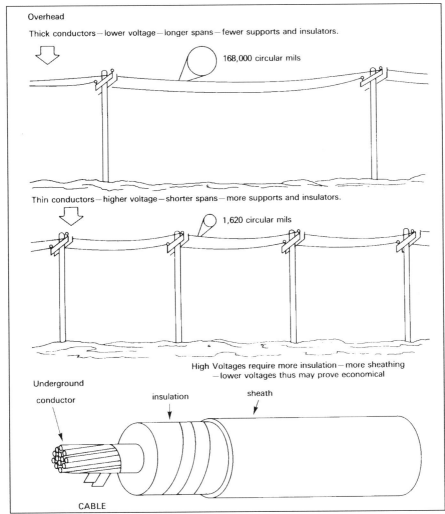

Fig. 1-11 Practical economics affect the size of a transmission line

Determining transmission voltages is a matter that requires careful study. Engineers "cost out" the systems, employing several generally standard voltages (and standard materials and equipment). For example, they might calculate the annual costs for each of the following systems: 69 kV, 138 kV, and 230 kV. Approximate costs of conductors and necessary equipment, insulators, supports, rights-of-way, labor costs, and associated expenses involved in construction and installation are carefully evaluated.

The annual carrying charges are calculated, to which are added the estimated annual maintenance expenses and the value of the losses for each of the systems under study. The lowest sum of these annual costs usually determines the selection of the voltage. The lowest overall annual expense will generally be found to occur when the annual carrying charges are approximately equal to the annual cost of the electrical losses incurred; this is known as Kelvin's Law.

In making these determinations, the future increased demand is also taken into consideration. The increase over a period of time is estimated and is included in the study. Again, economic studies determine the extent of the future time period to be considered, at the end of which facilities would need to be reinforced or replaced.

In the calculations involved in arriving at the selection of the transmission voltage, the construction, maintenance, and operation of the projected facilities are subject to the overriding considerations of safety. In general, the utility standards are higher than those recommended by the National Electric Safety Code and those imposed by local and governmental regulations.

Review

- The system of supplying electricity to a community includes facilities for production (generating plants), for wholesale or bulk delivery (transmission), and for local retailing (distribution).

- Transmission circuits may consist of open wires on towers or poles in outlying zones, or of cables installed in ducts underground or buried directly in the ground in more densely populated areas. Transmission line installations can provide for an interchange of power between two or more neighboring utility companies to their mutual advantage. Power may also be "wheeled" to non-adjacent companies via the facilities of the intervening companies. Such interconnections between utility companies are often referred to as a grid or pool.

- Transmission lines transport relatively large amounts of electric power over relatively long distances. Power may range from less than 10,000 kW and a few miles to more than 3 million kW and many hundreds of miles. The voltages at which these lines operate range from a few thousand volts to more than 765 kV.

- The voltage of a transmission line is usually determined by economics. Higher voltages are generally desirable since smaller, less-expensive conductors may be used. However, for overhead lines, the mechanical strength of the conductor may limit the span length, requiring a greater number of poles, insulators, etc. The annual costs of several standard voltage lines are evaluated; the lowest annual cost usually determines the selection of voltage.

- Transmission lines between substations and local or regional load centers are sometimes called *subtransmission lines*. They generally operate at voltages of 138 kV volts or less.

- Lines operating at 500 kV or more are generally referred to as extra-high-voltage (EHV) lines.

- Overhead transmission lines are more suited to open country areas, where rights-of-way of ample width and reasonably straight lines are available. Appearance also is not always of paramount importance.

- While overhead lines are exposed to the hazards of vandalism and weather, faults are usually relatively easy to find and repair. Conversely, underground lines are relatively free from harmful exposure. However, the faults that do occur are relatively hard (and time-consuming) to find and repair.

Study Questions

1. What is the function of transmission lines or systems?
2. What factors influence the voltage selected for a transmission line?

3. How does the load-carrying capability of a transmission line relate to the voltage at which the line operates?
4. What are some typical standard voltages of transmission lines?
5. What are subtransmission lines? At what voltages do they operate?
6. How do generating plants affect the transmission system?
7. What are the advantages of overhead-constructed transmission lines?
8. What are the disadvantages of overhead-constructed transmission lines?
9. What are the advantages of underground-constructed transmission lines?
10. What are the disadvantages of underground-constructed transmission lines?

2

Overhead Construction

The several elements constituting an overhead transmission line, together with pertinent information concerning their purpose, characteristics, and methods of handling, are covered in this chapter.

Poles and Towers

Structures for supporting the overhead conductors are broadly classified as poles (see Fig. 2–1 and Fig. 2–2) or towers (see Fig. 2-3 and Fig. 2–4).

Single pole supports may be spaced a few feet apart or more than 100 feet (ft) apart. They may be made of natural wood: southern yellow pine, western red cedar, douglas fir, larch, and other species, reflecting the availability of suitable timber in the several geographic areas of the country. They can also be made of imported wallaba where extra strength is required. These may be chemically treated, round-shaped tapered poles or square-shaped poles. Poles may also be of hollow, tapered, tubular design, made of steel or aluminum. They may also be built up of flat metal members, latticed together into a variety of cross sections, from squares to duodecagons (12-sided).

Wood poles are set directly in the ground. Metal poles may be embedded directly in earth or concrete, or secured to bolts embedded in a concrete

Fig. 2-1 Wood pole

base. Wood and metal poles may be combined as members of a structure into A-frames, H-frames, and sometimes into V- or Y-type transmission line supports capable of carrying longer spans.

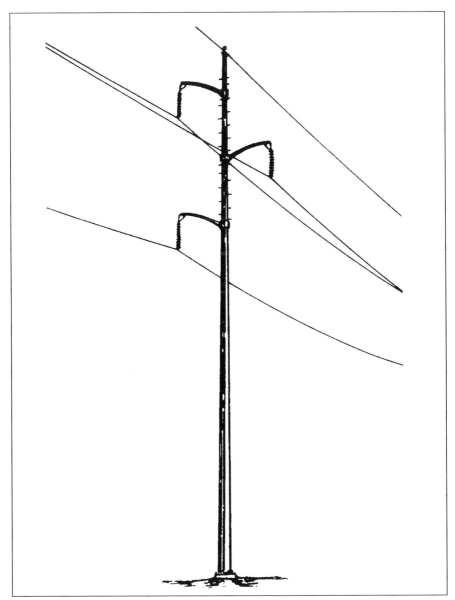

Fig. 2–2 Hollow, tapered, tubular pole

Reinforced concrete, hollow, round, or square poles have long life, structural strength, freedom from weathering, good appearance, and low maintenance costs. They are, however, relatively expensive in first cost and extremely heavy, making their handling difficult.

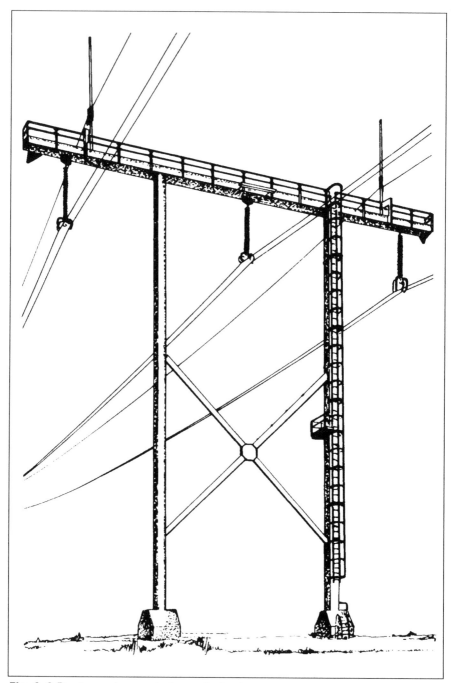

Fig. 2–3 Prestressed concrete transmission tower

Fig. 2–4 Steel lattice tower

Tower Types

Common long-span construction consists of structures suitably lat-ticed or braced and made of galvanized steel or aluminum. They are mount-ed on wide rectangular bases and sometimes are referred to as towers (see Table 2-1). Their footings may be of reinforced concrete or grillages made of assembled steel beams. There are several types of towers:

- **Tangent towers.** The conductors supported are essentially a straight line
- **Small-angle towers.** The conductors supported change direction slightly, perhaps 5°–10°

Table 2-1 Typical Pole Setting Depths

Pole Length (ft)	Approximate Setting Depth	
	All Solid Rock (ft)	Firm Soil (ft)*
35	4.5	6.0
40	5.0	6.0
45	5.5	6.5
50	6.0	7.0
55	6.0	7.5
60	6.5	8.0
65	7.0	8.5
70	7.0	9.0
75	7.0	9.5

*Based on approximately 10 percent of pole length plus 2 feet.

- **Medium-angle towers.** These have support changes of 20°–30°
- **Heavy-angle towers.** These accommodate sharper turns
- **Dead-end towers.** They support the entire pull of conductors that terminate on them and are located at anchor points in a line, at certain angles in the line, or at points of take-off from the line

Structures may be of self-supporting latticed (tower) construction or guyed. These latter may be modified **A**, **H**, **V**, **Y**, or guyed-mast types, designed to give certain flexibility to the structure (see Fig. 2–4). Towers or structures are usually referred to by type, voltage, and number of circuits, such as: H-frame, tangent, 138-kV, and double-circuit tower. In many cases, the transmission line will contain two or more types of supporting structures.

Guy Wires and Anchors

Guy wires (guys) and anchors are installed where lines terminate (referred to as dead-ends), at angles, on long spans where pole strength may be exceeded, and at points of excessive unbalanced conductor tension. Guying for line protection is installed at various points in the line to limit damage should storms or other violent action cause the line to fail in somewhat of a domino fashion.

While practically all wood or metal pole transmission structures have guy and anchor installations associated with them, occasionally, some self-sustaining towers may also require their installation. One example is at 90° turns in the line (see Figs. 2-5, 2–6, 2–7, and 2–8). Both guy wires and

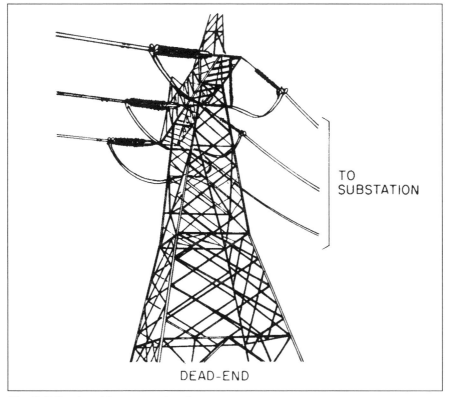

TO
SUBSTATION

DEAD-END

Fig. 2-5 Dead-end tower construction

anchors used in transmission lines are generally heavier than those for distribution lines.

Distribution Underbuild

In some instances, transmission poles may also carry distribution circuits, generally in a lower position on the pole. Design of such common lines requires extreme care to protect the distribution lines from the effects of the higher voltage lines and to provide safe working conditions for the workers. Additional guying may be necessary to compensate for the effect of such underbuild on transmission structures (see Fig. 2-9).

Design Factors

The choice of supporting structures is influenced by many factors that, considered together, result in the greatest economy. The voltage of the

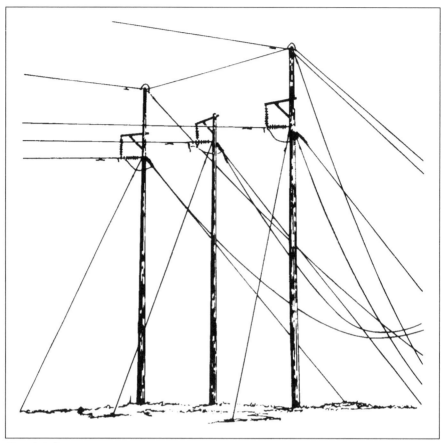

Fig. 2-6 Guyed structures

line or lines to be carried, both present and future, determine the spacing of the conductors and the length of the string of insulators required. This may be subject to modification to provide proper space and conditions for the workers. Span length in turn is affected by the clearances required for safety (see Table 2-2).

These minimums are usually specified by the National Electric Safety Code (NESC), whose provisions are established and revised periodically by the Institute of Electrical and Electronics Engineers (IEEE). They generally reflect the latest practices; reference should therefore be made to the latest code revision. They are also influenced by local ordinances, and by the allowable difference in sags of the conductors under summer and winter conditions.

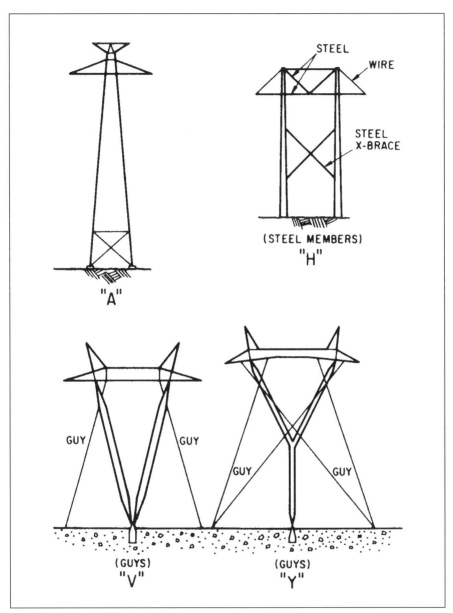

Fig. 2-7 Wood and metal pole constructions

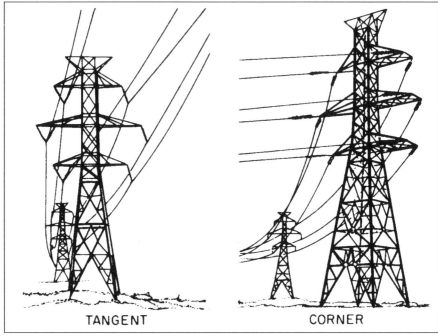

TANGENT   CORNER

Fig. 2–8 Tangent and corner tower constructions

The size, type, and material of the conductor also affect span length. Aluminum conductors generally have a greater summer sags than copper conductors. Consequently, they require higher supports to achieve the same minimum clearance above ground and also to keep the conductors from swinging into each other (see Table 2-3). Appearance, the character of the terrain (whether plain or hilly), climatic conditions, the nature of the soil, and transportation facilities have to be evaluated. The structure itself— wood, metal, or concrete—is also a determining factor.

The first cost of such supports is not the only consideration; the costs of operation and maintenance are also important. The longer the spans, the fewer will be the number of points of support. There will also be fewer insulators necessary requiring possible replacement, fewer structures requiring painting, and fewer poles requiring replacement. All of these factors may vary with the time of installation, location, local conditions, prevailing regulations, and with plans for the future. (For example, there might be an addition of a second line several years after the initial installation.) However, the final decision is usually the result of many judgments and compromises.

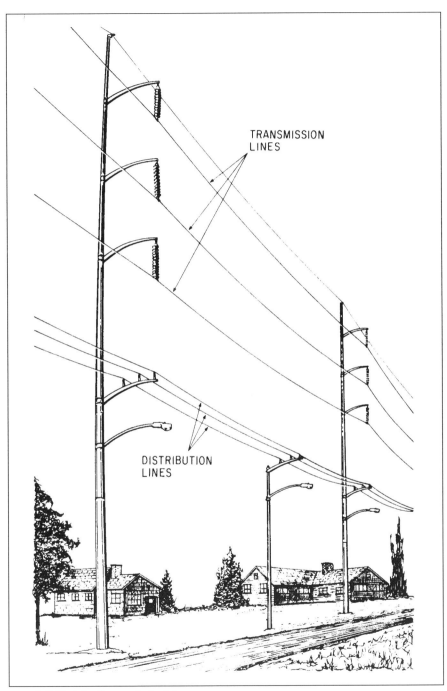

TRANSMISSION
LINES

DISTRIBUTION
LINES

Fig. 2-9 Distribution underbuild

Table 2-2 Recommended Minimum Vertical Clearance of Conductors (in ft) Above Ground or Rails

(120°F, No Wind, Final Sag. Conductor Design Tension 50% of Ultimate Strength)

Nature of ground or rails underneath	34.5 kV	46 kV	69 kV	115 kV	138 kV	161 kV
Track rails of railroads	32	32	33	35	35	36
Public streets & highways	25	25	26	28	28	29
Areas accessible to pedestrians only	19	19	20	22	23	24
Cultivated fields†	20	20	21	23	24	25
Along roads in rural districts	22	22	23	25	26	27
Natures of wires crossed over						
Communication lines	8	8	9	11	11	12
Supply lines up to 50,000 volts*	6	6	7	9	9	9

Sag should allow one foot greater clearance than shown above. Conductors smaller than 1/0 ACSR may require additional clearance.

†The NESC does not specify the clearances that should be maintained across cultivated fields.

Local conditions and regulations may call for deviations from these recommended minimum clearances.

*For lower conductor at its initial unload sag.

Always refer to the latest edition of the Code.

Table 2-3 Clearances in Any Direction from Line Conductors to Supports and Guy Wires Attached to the Same Support

	Suspension Insulators			Minimum Clearance		
Rated Line Voltage (kV)	No. Units*	Weight (lb)	Normal Clearance to Support	NESC	To Support	To Guy Wires
34.5	3	38	1'7"	0' 10"	1' 0"	1' 5"
46	3	38	1'7"	1' 1'	1' 0'	1' 9"
69	4	48	2' 1"	1' 6"	1' 3"	2' 7"
115	7	78	3' 6"	2' 6"	2' 2'	4' 1"
138	8	88	4' 2"	3' 0"	2' 9"	4' 10"
161	10	108	5' 0"	3' 6"	3' 6"	5' 7"
230	12-14	138	6' 10"	4' 11"	5' 0"	7' 11"

* For average conditions tangent structure.

Computers can be used to make these calculations, varying each and every factor as desired. Further, they can do this in the time that engineers previously calculated by hand only a few choices, using only a few important variables. In general, the calculated results will depend on the estimates of cost, depreciation allowances, and power to be transmitted.

Cross-arms

Cross-arms for towers are generally of galvanized steel or aluminum; wood is used for pole and A- or H-frame construction. Laminated wood cross-arms and arms made of epoxy synthetics have also been used (see Fig. 2-10). Since these materials are insulators, the number of insulators in a string may be reduced. The other advantages are better appearance and better working conditions because of the completely insulated arm.

Conductors

The size, type, and material of conductors are affected by the amount of power desired to be transported and the distance it is to be transported. Both factors largely influence the voltage at which the line is to operate.

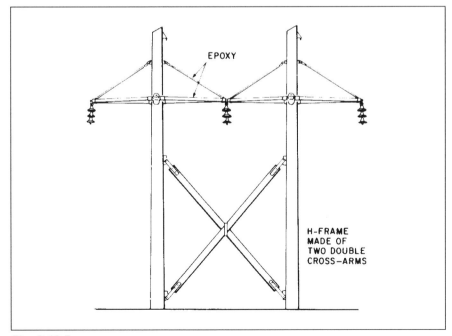

Fig. 2-10 Epoxy cross-arms

Theoretical considerations indicate that the most economic conductor will be the one whose annual carrying charges equal the annual cost of power losses in the conductor. (Annual carrying charges include interest on investment, operation and maintenance costs, taxes, and insurance.)

However, the drop in voltage and the heating of the line may be excessive. In such cases, the economy of the smaller conductor line may have to give way to one with larger conductors. This may be necessary in order to produce tolerable voltage losses and heating conditions (see Fig. 2-11).

The size of conductor may have to meet mechanical considerations and electrical losses because of the phenomenon known as "skin effect" and corona (explained later in this chapter). Too thin a conductor may require additional supports due to its lack of mechanical strength. On the other hand, large conductors are difficult to handle and may impose larger-than-desirable strains on insulators. One remedy is to subdivide the circuits into two or more circuits, which may add other advantages in reliability.

Self-inductance and Skin Effect

When a conductor carries an alternating current, it produces around itself a magnetic field whose intensity varies with the changing values of the alternating current (see Fig. 2-12). This changing magnetic field cuts the conductor, inducing within it voltages that tend to retard the flow of current normally flowing in the conductor. This phenomenon is known as self-inductance or self-reactance.

Within a conductor, this effect is more pronounced toward the center of the conductor. Hence, current flowing within that conductor will tend to flow more easily (and consequently in greater part) near the surface of the conductor. This is known as "skin effect" and is more pronounced as the operating voltages increase. Where the conductor is stranded, the outer strands carry more current per strand than do the inner strands.

Corona Discharge

When a conductor carries a voltage exceeding a certain critical value, a halo-like glow, known as a "corona," will appear on the surface of the conductor (see Fig. 2-13). The amount of corona discharge depends on the diameter of the conductor, the condition of the adjacent atmosphere, other nearby conductors, and condition of the surface of the conductor, such as dirt or roughness.

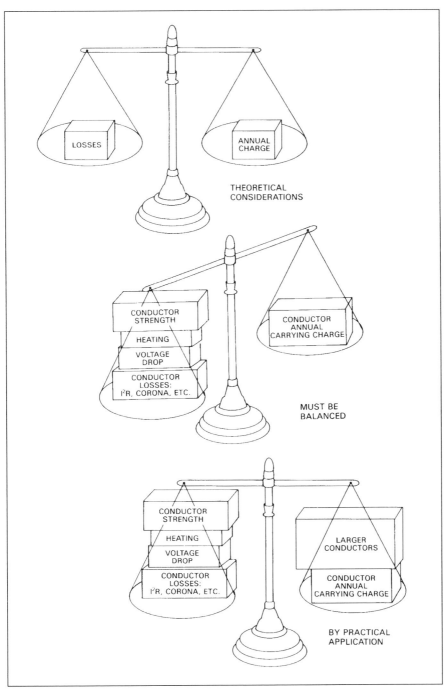

Fig. 2-11 Steps in selecting the right conductors

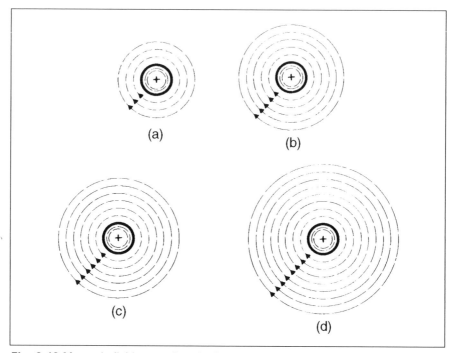

Fig. 2-12 Magnetic field expanding (**a, b, c, d**) and contracting (d, c, b, a) about a conductor

The luminous effect of the corona discharge is from a discharge of electrical energy from the conductor into the atmosphere, where it is dissipated and represents a loss of power (I^2R). If the distance between this conductor and other nearby conductors or structures is comparatively small, a sparkover may occur, triggering a short circuit at this point with consequent interruption or damage to the line.

Corona discharge and skin effect may cause interference on communication circuits that parallel the transmission lines. Further, corona discharge may also interfere with local radio and television broadcasting.

Hollow Conductors

To lessen the impact of skin effect and corona, expanded conductors having hollow or partially hollow cores have been developed. Such conductors eliminate the center part of the conductor, which is not fully used in carrying current because of the skin effect. Further, because a conductor of larger outside diameter results, the tendency for corona to appear is decreased.

Fig. 2-13 Corona effect

A wire or conductor in which electricity may be flowing is surrounded not only by a magnetic field, but also by an electrostatic field. Electrostatic fields generally form in uniform patterns around a straight conductor and are also conductors of electricity. These patterns tend to become concentrated at points where the conductor is bent or at its edges or ends; the conducting tendency is also increased at these points.

Since corona discharge tends to appear at points where the electrostatic fields around a conductor are concentrated, sharp bends, corners, and points should be avoided. Corners or bends should be made gradual and smooth. Corona discharges may be greater, and actually observed, during

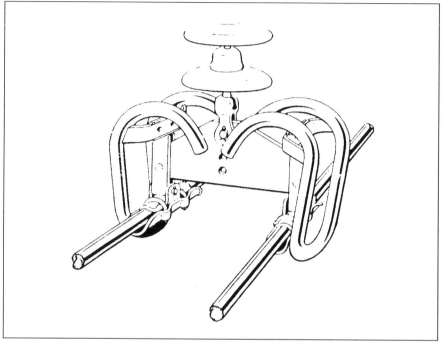

Fig. 2-14 Corona shield

period of rain. The raindrops accumulating on a conductor essentially change its shape; the clinging drops create relatively sharp pips on the conductor, encouraging the formation of corona at these points.

Corona Shields

To prevent corona flashover damage to insulators, particularly during inclement weather, shields (see Fig. 2-14) or rings (see Fig. 2-15) are provided at both the conductor and the supporting end of the insulators. This furnishes a path for the flashover away from the surface of the insulators. The insulators are thereby protected from the shattering effect of the current that flows during the flashover.

Conductor Materials

Conductors can be made from various types of materials (see Fig. 2-16). High conductivity (lesser resistance to the flow of electricity), great strength, and elasticity favor hard-drawn copper for use as a conductor.

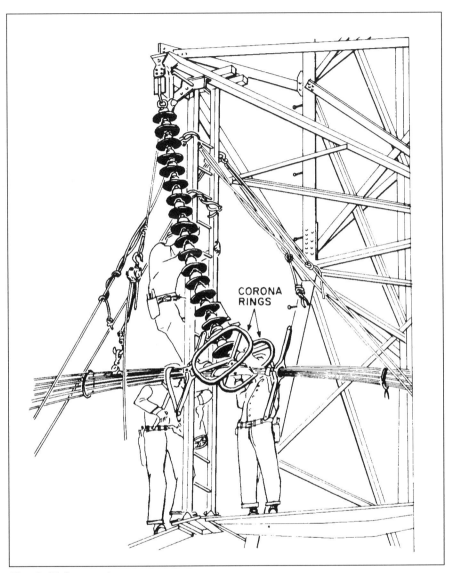

CORONA
RINGS

Fig. 2-15 Corona rings

However, aluminum has an advantage because of the requirement for a large diameter to avoid corona. This, together with price advantages, makes it almost universally used in high-voltage transmission lines.

The lower conductivity of aluminum, about 60% that of copper, results in a conductor having a diameter of 1.26 times that of a copper con-

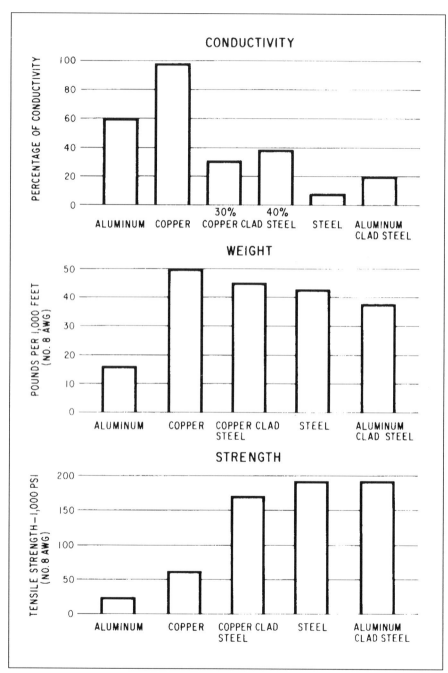

Fig. 2-16 Comparison of various conductor materials according to conductivity, weight, and strength

ductor of equal resistance. In addition, its lower weight makes it somewhat easier to handle. However, the lower strength of aluminum, about 70% that of copper, makes it less suitable for long spans. To overcome this, aluminum cable, steel reinforced (ACSR) was developed. In this type of cable, aluminum conductors are wrapped around a steel cable that provides the strength necessary for longer spans. Later designs have replaced the inner steel cable with one made of a high-strength aluminum alloy.

For shorter lines designed to carry smaller amounts of power, conductors made of copper-clad steel and aluminum-clad steel provide mechanical strength. They also furnish the relatively large diameters needed to mitigate skin and corona effects. For higher capacities, several of these clad wires may be used as a core, around which copper or aluminum strands may be laid to form a reinforced conductor.

Steel conductors are occasionally used for extra-long spans, such as river crossings, where very high mechanical strengths may be required. The self-inductance is greater because of the magnetic properties of steel. As a result, the skin effect may be great enough to require a larger cross section for the steel conductor than for equivalent copper or aluminum conductors. The larger diameter serves to increase the value of critical voltage at which corona forms.

To minimize corona effects, the overall diameters of ACSR (and other) conductors have been expanded by introducing intermediate suitable fibrous material. Corona effects are also minimized by forming air pockets between the outer aluminum strands and the inner steel core. Another means of minimizing corona losses is to replace the one conductor with a "bundle" of two, three, or four conductors. These are held in place by spacers and separated from each other by suitable distances up to about 18 in. Less expensive conductors may be used with this construction, though greater ice and wind loads may be experienced, and greater sag for a given span may result.

Conductors for high-voltage transmission lines are handled very carefully to keep them from scraping on the ground during installation. Often special tension-stringing equipment is used that eliminates scratches or sharp edges on the conductor, which may cause corona discharges. Aluminum is more subject to chemical attack and corrosion from substances found on the ground and in the atmosphere than is copper.

Natural Hazards

Overhead transmission lines are exposed to attacks by the forces of nature. These include not only rain, which affects the corona discharge on conductors, but also wind and ice loading of conductors. Provision must therefore be made in the design of supporting structures for the effects of ice, wind, lightning, and other ravages of nature.

Aeolian Vibrations

Overhead conductors of transmission lines in areas where span lengths are relatively long (usually in open country) are subject to vibrations and movement produced by wind.

One type, known as aeolian vibrations (see Fig. 2-17), is a regular high-frequency oscillation caused by the eddies behind a conductor pro-

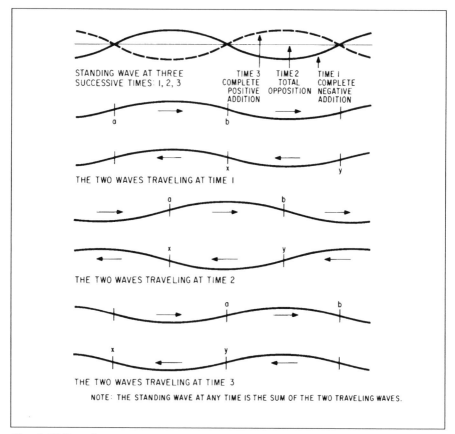

Fig. 2-17 Standing waves on a conductor because of aeolian vibrations

duced by wind. The frequency of vibration depends on the size of the conductor and wind velocity. To lessen the effect of these vibrations, dampers (masses of metal) are installed on the conductors at node points (see Fig. 2-18).

Generally, true node points are impossible to be determined because of the varying factors producing the vibrations. However, the dampers usually are located near the towers at points calculated to give as much dampening as is practical. Armor rods are also installed in aluminum conductors at the insulator clamps to reduce the wearing effect of the vibrations in such conductors.

A self-damping conductor is designed to reduce the aeolian vibration effect. This self-damping effect is achieved by making the shapes of the conductor strands different. The outer strands are trapezoidal, while the inner ones are round (see Fig. 2-19). Relatively large clearances between these conductors permit motion to take place within the conductor layers, which tends to break up the vibration of the conductor caused by wind blowing against and around it. With this conductor design, appropriate splicing and terminating materials and procedures are necessary.

Galloping or Dancing Conductors

Another type of vibration, much more severe, is known as "galloping" or "dancing" conductors. The cause for this occurrence is not always known. However, ice forming on the conductors during freezing rain or sleet storms may sometimes assume the approximate shape of an airfoil. Wind blowing against it may cause the conductor to be lifted appreciably until it reaches a

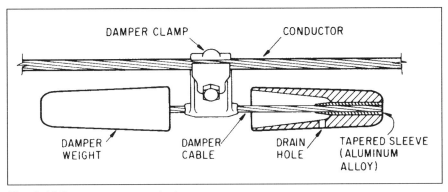

Fig. 2-18 Dampers on a conductor

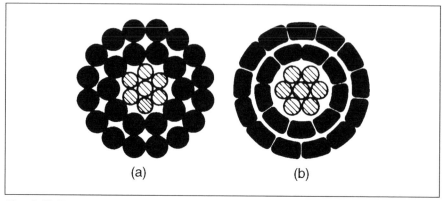

Fig. 2-19 Conventional ACSR (**a**) and the self-damping conductor (SDC)(**b**)

point where the conductor falls or is blown downward. Such a condition imposes extremely severe strains on both conductors and supporting structures and may cause them to fail.

Further, the increased sag of the conductors because of the ice load coupled with the nonrhythmic swaying may cause them to whip together. This could result in flashovers and short circuits, which may trip the circuits or burn down the conductors. Little can be done to correct this condition, which is fortunately rare, except to attempt to melt the ice from the conductors. This is sometimes done by overloading a circuit deliberately, either by transferring loads from other circuits or by connecting a "phantom" load to it (see glossary for definitions of these terms). Such overloads cause the conductors to heat and thus melt away the ice.

This practice of overloading has been successful in many instances, though sometimes uneven heating and dissipation of heat cause only portions of the conductors to become free of ice. If the circuits can be taken out of service, this is sometimes done, and the conductors are allowed to whip together with no danger of short-circuit and burndown.

Connectors and Splices

Mechanical connectors are almost universally used in splicing conductors. These may sometimes consist of sleeves or yokes that hold the conductors together by bolts. Sometimes the ends of the conductors are inserted into a sleeve, which is then crimped hydraulically. These are referred to as compression connectors.

These latter often have the indentations filled with solder, and the whole splice is polished so that the tendency for corona to form about the splice is reduced. For ACSR conductors, two sleeves are used, an inner steel sleeve fitting over the steel core only, and an outer sleeve of aluminum fitting over the entire conductor (see Fig. 2-20). For aluminum reinforced aluminum conductors, only one overall outer sleeve is used. Mechanical or hydraulic compressors are used to create the indentations on both the steel and aluminum sleeves that grip the conductors. Such indentations are usu-

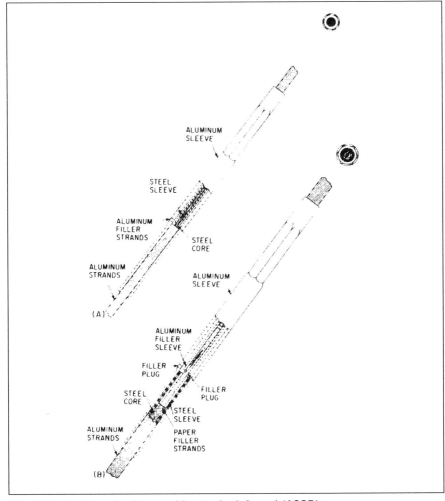

Fig. 2-20 Joint for aluminum cable, steel reinforced (ACSR)

ally started at the center of sleeve and proceed toward each end; the length of the sleeve may increase during compression.

Heat Detection

Splices or connections in the conductor are often sources of trouble if improperly made or oxidized. A loose connection or broken strands may create hot spots in the line, which may cause failure. Usually, these are difficult to discover. A bolometer (a device used for measuring heat at a distance) can be equipped with a television camera and mounted on a truck or aircraft. Used in patrolling the line, it can successfully ferret out these potential sources of trouble before failure occurs.

Infrared imaging has been advanced as a diagnostic tool for coronas and heat detection. However, while coronas produce some heat, it may not be detectable by infrared when compared to the heat of a conductor. Coronas usually are detected as static on radio receivers.

Insulators

Insulators used for transmission lines are both of the suspension and pin or post types. In general, for tower lines and A-, H-, V-, or Y- (wood or metal) frame construction, standard insulator discs 10 in. in diameter and spaced 5-1/4 in. apart are used. The number required depends on the voltage of the line. Standards adapted by the Edison Electric Institute and National Electrical Manufacturers Association specify minimum numbers of discs in a string for certain line voltages. For instance, some disc specifications are: 4 for 69 kV, 7 for 138 kV, 12 for 230 kV, and 19 for 345 kV. Extra discs are added to provide safety factors. They are also used to allow for contamination and damage and to compensate for the effect of altitude and metal structures (as compared to wood) on insulation requirements.

Suspension Insulators

The higher the voltage, the more insulation value is needed. Transmission lines operate at high voltages, such as 69 kV, for example. At these voltages the pin-type insulator becomes bulky and cumbersome. Besides, the pin that must hold it would have to be inordinately long and large. To meet the problem of insulators for these high voltages, the suspension insulator is used (see Fig. 2-21).

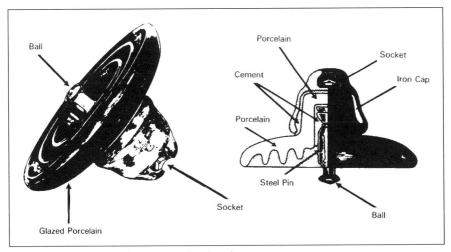

Fig. 2-21 Ball-and-socket suspension insulator

The suspension insulator hangs from the cross-arm, as opposed to the pin insulator, which sits on top of it. The line conductor is attached to its lower end. Because there is no pin problem, any distance between the cross-arm and the conductor may be obtained just by adding more insulators to the "string."

The entire unit of suspension insulators is called a string. How many insulators this string consists of depends on the voltage, prevailing weather conditions, the type of transmission construction, and the size of insulator used. It is important to note that in a string of suspension insulators, one or more insulators can be replaced without replacing the whole string.

Insulator Assemblies—Disc Type

Disc insulators are usually made of porcelain and designed so the porcelain is in compression (see Fig. 2-22). Though they may be made in various sizes and shapes for specific purposes, the standard type disc mentioned above has received wide acceptance. Glass and some synthetic (epoxy, polymer) types are also in use. (See Appendix H, "Porcelain vs. Polymer Insulation". Where additional mechanical strength is required, as in very long spans, two or more strings of insulators may be paralleled to provide the additional strength.

The string of insulators may support a conductor vertically from a cross-arm or structure. Alternately, it may support a conductor horizontally

Fig. 2-22 Typical string suspension insulator

(a string each direction) where strains are unusually great, as at turning points in the line (see Fig. 2-23). (Insulator strings can also be used in line applications with no strain as shown in Fig. 2–24.)

"Strain" or "dead-end" insulator strings are also used on a long line to divide it into several sections. Consequently, under unusual conditions such as storms, if a conductor fails, damage will be restricted to a small section. The unbalance caused by such a conductor failure may affect adjacent spans in a domino-like effect. The installation of these dead-end strings prevents the entire line from being affected.

Pin and Post Insulators

Pin insulators (see Fig. 2-25 and Fig. 2–26) and post insulators (see

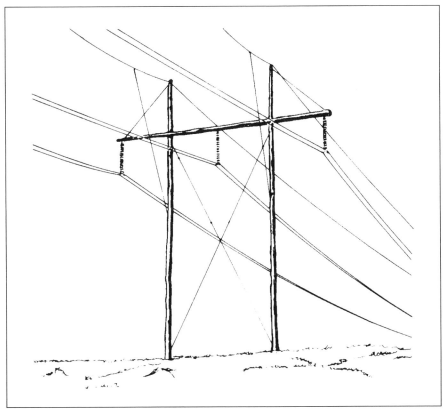

Fig. 2-23 Disc insulators—strain application

Fig. 2-27 and Fig. 2–28) are generally confined to transmission lines of lower voltages, usually up to 69 kV. These may be mounted on cross-arms where, since the conductor is mounted on the top of such insulators, the pole height can be reduced by that amount. Post-type insulators may be mounted horizontally on poles in a so-called "armless" construction, which has the advantages of better appearance and narrower right-of-way requirements.

Combinations of string insulators and horizontally mounted posts or insulated struts are used to control the position of the conductor, especially at angles in the line. Two strings of insulators, mounted from a cross-arm in a V position, are also used to hold a conductor in place away from towers or wood frame construction. These are referred to as V-strings (see Fig. 2-29 and Fig. 2–30).

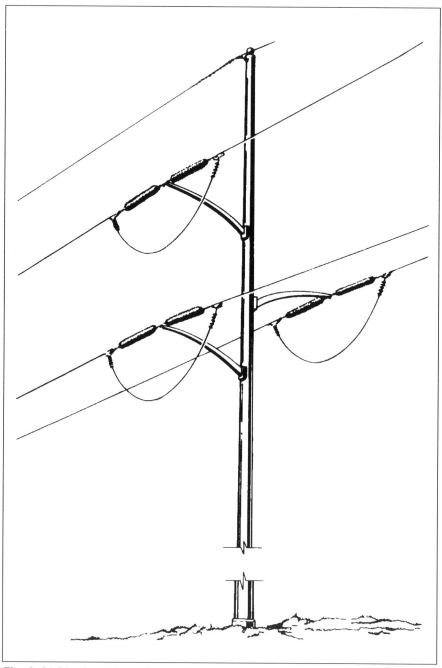

Fig. 2–24 Disc insulators—line application (no strain)

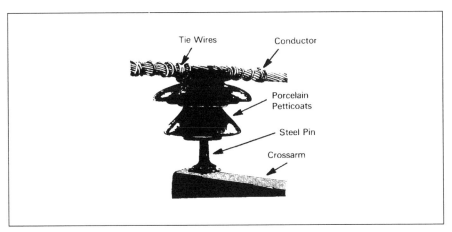

Fig. 2-25 Mounted pin-type insulator

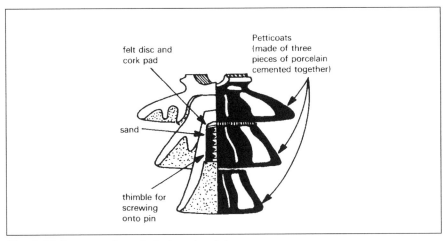

Fig. 2-26 Cutaway view of pin-type insulator

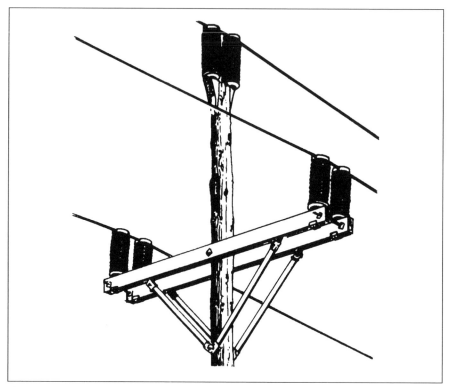

Fig. 2-27 Post-type insulators using cross-arms

Fittings

Fittings associated with the insulators are made of galvanized steel, malleable iron, or aluminum. The first two are used where strength requirements are high, the last for corrosion resistance and lessened corona effect because of the greater smoothness that can be attained. Aluminum-coated steel fittings are also available.

Lightning

When lightning strikes at or near a transmission line, voltages are created in those lines greater than those at which the line normally operates. These "induced" voltages may be destructive if allowed to flashover to other lines or structures. Means are provided to "bleed" off these surges harmlessly. Arcing rings, mounted at one or both ends of the insulator string, permit the high-voltage surge to flashover between the ring and the structure or

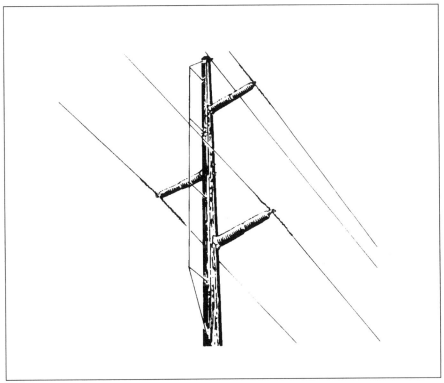

Fig. 2–28 Post-type insulators using armless construction

between these rings, away from the porcelain insulators. The energy is passed over to the ground, where it is dissipated (see Fig. 2-31).

Lightning or Surge Arresters

Lightning or surge arresters, usually a series of air gaps or special semiconducting materials, also are used to drain away the voltage surges. These are generally placed at the terminals of the lines to protect transformers, circuit breakers, and other equipment (at the substation). They also may be placed along the lines at strategic locations.

The elementary lightning arrester (see Fig. 2-32) consists of an air gap (horn air gap) in series with a resistive element. The overload voltage surge causes a spark that jumps across the air gap, passes through the resistive element (silicon-carbide, for example). The resistive element is usually a mate-

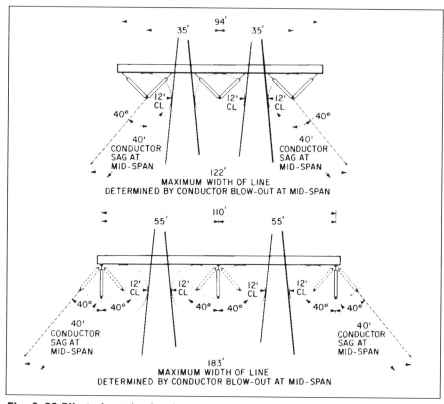

Fig. 2–29 Effect of v-string insulator configuration on the right-of-way requirement

rial that allows a low-resistance path for the high-voltage surge, but presents a high resistance to the flow of line energy.

There are many different types of arresters, but they all have one principle in common. There is always an air gap in series with a resistive element. Whatever the resistive (or valve) element is made of, it must act as a conductor for high-energy surges and also as an insulator toward the line energy. In other words, the lightning or surge arrester leads off only the surge energy. Afterwards, there is no chance of the operating line voltage being led into the ground.

The valve element in a resistor arrester (see Fig. 2-33) consists of ceramic-like discs that act as conductors under high-voltage surges and present a high resistance to the line energy. In the arrester shown in Figure 2–34, the lightning current passes through a series of bypass gaps to the main gap

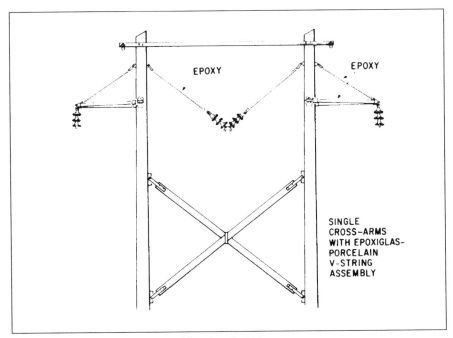

EPOXY   EPOXY

SINGLE
CROSS-ARMS
WITH EPOXIGLAS-
PORCELAIN
V-STRING
ASSEMBLY

Fig. 2–30 V-string insulators and insulated struts

and an element to ground. If the line energy attempts to follow the lightning energy, that line energy is made to flow through a series of coils that create a magnetic field. The magnetic field is strong enough to extinguish the arc of the lightning discharge (see Fig. 2–35). This extinguishing action is so rapid that it takes place in less than 1/2 cycle of the line energy.

Shield or Ground Wires

Shielding these lines by means of a wire above them, which is directly connected to ground, is another form of protection from lightning. These wires will drain off the voltage surges from direct or nearby strikes of lightning (see Fig. 2-36). Further, they cause the air adjacent to the lines to be drained of static electricity that sometimes serves to "attract" lightning discharges. This acts much the same as the ordinary lightning rod and may be thought of as a continuous lightning rod, protecting the lines.

One or more wires may be mounted on the structures supporting the transmission line (see Fig. 2-37), depending on the type of structure and number of circuits. It has been found that the installation of the overhead

Fig. 2–31 Arcing rings on the insulator

shield wires (at an angle of 30° or less from vertical) covering the lines generally results in the greatest protection of these lines. Wire for this purpose is generally of galvanized steel of small diameter. However, copper-clad steel is sometimes used to reduce electrical resistance while maintaining mechanical strength.

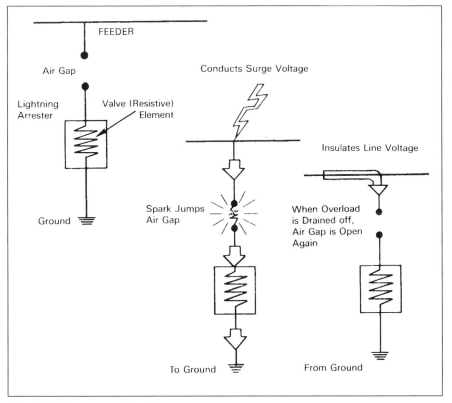

Fig. 2-32 Elementary lightning arrester

Counterpoise

To help this system drain off the lightning discharges, its resistance to the ground is kept low. This is further improved by burying rods or wire radially around the bases of the towers or supports, or burying a wire between them. This is known as a "counterpoise" (see Fig. 2-38), and its function is to dissipate the discharge over a larger area of ground, thereby lessening the resistance to the flow of these discharges. It is desirable, therefore, for the tower footing resistance to be made as low as possible; a value of 5 ohms (Ω) or lower is generally sought. The question now arises as to how the value of tower footing resistance is measured and how it may be reduced.

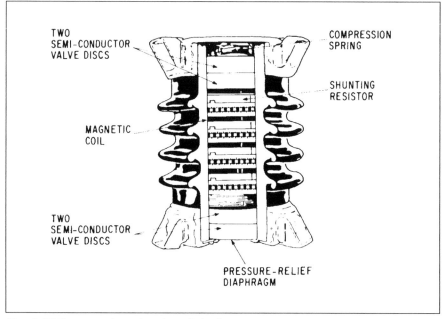

Fig. 2-33 Cutaway view of resistor lightning arrester

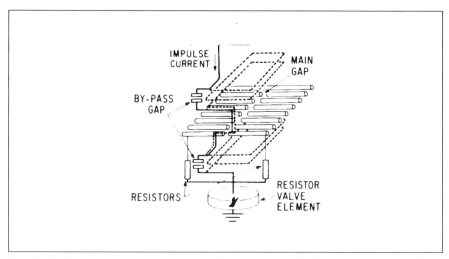

Fig. 2–34 Diagram of resistor arrester showing path of lightning current in solid line

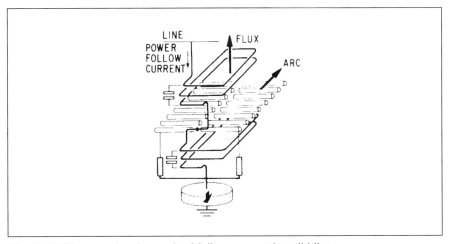

Fig. 2–35 Diagram showing path of follow current in solid line

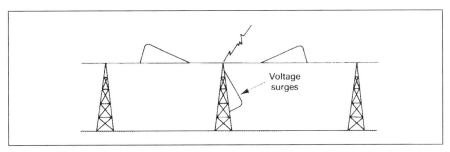

Fig. 2-36 Transmission-line ground wire struck by lightning

Methods of Measurement

There are three general methods used to measure the ohmic resistance from the tower to earth.

Fall of Potential Method

In Figure 2–39, a known current, measured by the ammeter, flows in the ground between the tower and electrode B. The voltage drop, or fall of potential, between the tower and electrode A is measured with a voltmeter. The tower footing resistance (R_x) is then determined by the relation:

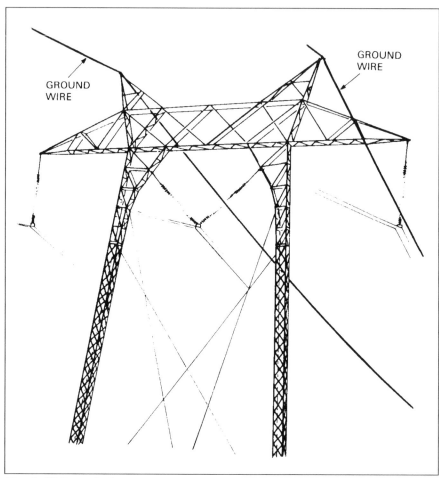

Fig. 2–37 Ground wires

$$\text{Footing Resistance } (R_x) = \frac{\text{Fall of potential}}{\text{Known current}}$$

The electrodes should be located sufficiently apart to avoid proximity effect on each other.

Three-Point Method

In Figure 2-40, three different resistance measurements are obtained: R_x in series with resistance of electrode R; R_x in series with the resistance of

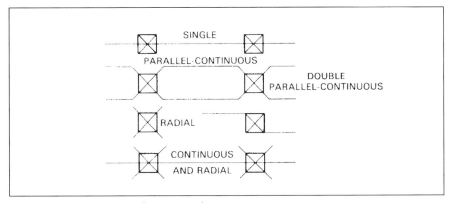

Fig. 2–38 Arrangement of counterpoises

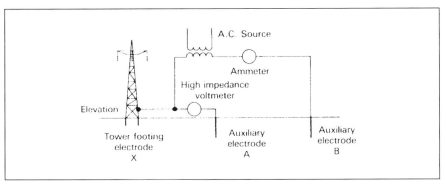

Fig. 2-39 Fall of potential method of measuring ohmic resistance

electrode R_B; and R_A in series with R_B. Three measurements are taken, each one in turn, by using the voltmeter-ammeter method of measuring resistance. The footing resistance (R_x) is then computed from the equation:

$$R_x = \frac{(R_x + R_A) + (R_x + R_B) - (R_A + R_B)}{2}$$

The electrode should have resistances of the same order of magnitude as that of the tower footing.

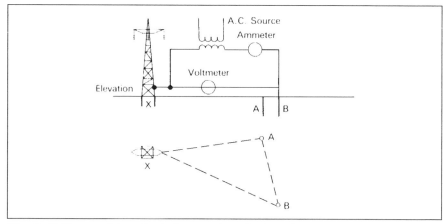

Fig. 2–40 Three-point method of measuring ohmic resistance

Ratio or Wheatstone Bridge Method

In Figure 2–41, $R_x + R_B$ is first measured with the Wheatstone bridge. (This is a laboratory measuring device based on a comparison of current and resistance ratios used to determine unknown quantities of resistance.) Then:

$$R_1 \text{ is to } (R_1 + R_2) \text{ as } R_x \text{ is to } (R_x + R_B)$$

or

$$\frac{R_1}{(R_1 + R_2)} = \frac{R_2}{(R_2 + R_B)}$$

but

$$R_1 + R_2 = R_2 + R_B$$

and

$$R_1 = R_2$$

$$R_2 = \frac{R_1(R_2 + R_B)}{(R_1 + R_2)}$$

or

$$R_x = \frac{(R_x + R_B) \, R_x}{(R_x + R_B)}$$

It should be realized that the resistance of an earth connection varies with many factors such as moisture, temperature, depth and diameter of electrodes, season of the year, and earth composition.

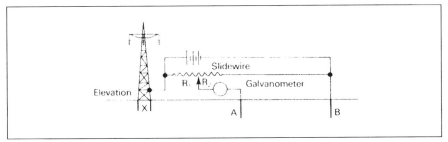

Fig. 2–41 Ratio or Wheatstone bridge method of measuring ohmic resistance

Methods of Reducing Tower Footing Resistance

Resistance of tower footings may be reduced by: installing additional conductor paths in the counterpoise; driving additional ground rods and connecting them to the counterpoise; or a combination of both.

In some regions, it is fairly easy to obtain tower-footing resistances of 5 Ω or less; in other regions, it may be more difficult. Soil resistivity is the most important factor in determining the resistance of tower footings. Some typical values are shown in Table 2-4.

Surges

Similar "surges" will occur in transmission lines from switching operations or from intermittent temporary short circuits. The first are created

Table 2-4 Typical Values of Soil Resistivity*

Soil	Resistivity Range
Clay, moist	14-30
Swampy ground	10-100
Humus and loam	30-50
Sand below ground water level	60-130
Sandstone	120-70,000
Broken stone mixed with loam	200-350
Limestone	200-4,000
Dry earth	1,000-4,000
Dense rock	5,000-10,000
Chemically pure water	250,000
Tap water	1,000-12,000
Rain water	800
Sea water	0.01-1.0
Polluted river water	1-5

*In ohms per cubic meter.

when a switch or circuit breaker is opened on interconnected lines, and the surge is created much as in a water system when a valve is suddenly closed. The latter may be caused by tree limbs, kites, or other foreign objects making temporary contacts. Rapid relaying is sometimes provided, which takes the lines out of service before damage to them or other equipment can occur.

As soon as the transient disturbance is removed, service is restored to them by automatic reclosing devices called "reclosers." Figure 2-42 is an oscillogram showing a typical example of a recloser operation. Notice that the first time it opens and closes, the action is instantaneous, requiring only 1.6 cycles. The second time the action is delayed to 2 cycles, the third time to 6, and the fourth time to 5-1/2 cycles. Then the recloser locks itself open, and a worker must correct the fault and manually close the mechanism.

The timing mechanism (see Fig. 2-43) that can effect this sort of action is a hydraulic system utilizing transformer oil as retarding fluid. With this device, the second, third, and fourth openings of the recloser can be set to any desired time span. To adjust the second and third openings, the time plate must be positioned. To designate whether the lockout should occur after two, three, or four operations, the cotter pin must be placed in the proper hole. To select the number of fast operations (one, two, or three), the roll pin must be positioned. If it is desirable to have all four operations be instantaneous, the timing plate must be removed.

Maintenance

When such disturbances occur, it is usually customary to patrol the line to look for places of fault. Flashover marks, pitted conductors, chipped

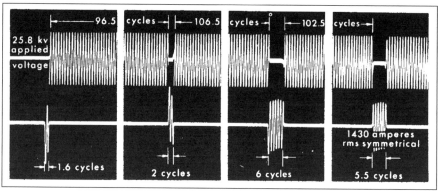

Fig. 2-42 Oscillogram showing automatic recloser operation

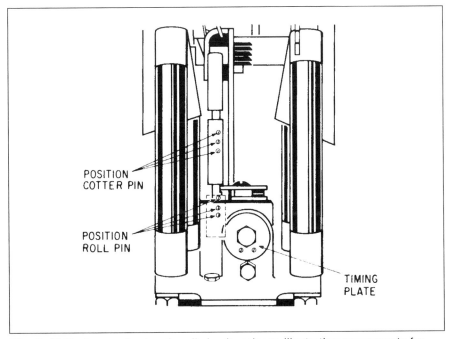

POSITION
COTTER PIN

POSITION
ROLL PIN

TIMING
PLATE

Fig. 2–43 Timing mechanism for oil circuit recloser, illustrating components for sequence adjustment

or stripped insulators, and fallen overhead ground or shield wires may be found. Maintenance and replacement of damaged units usually follow to prevent these relatively minor incidents from developing into major sources of trouble. Helicopters and aircraft have been used, as well as patrols by foot, horseback, and car.

Live-line Operation

Insulators, conductors, and other items may be handled while energized with live-line (hot stick) or bare-hand methods. Ingenious tools, mounted on the end of wood sticks, are used in live-line maintenance. The wood acts as insulation. The worker fastens the stick to a conductor, after first disconnecting it from the insulator or insulators. Lashing the stick to the pole or tower keeps the energized conductor a safe distance away while insulators are changed, structure repairs are made, or ground shield conductors are reinstalled.

More sophisticated procedures are used to replace connectors, or wrap reinforcing rods around pitted conductors or broken strands. Care must be taken to see that workers always have sufficient insulation or space between them and energized parts of the line. For additional mechanical safety, conductors and other energized items are usually supported by one or more hot sticks, or by a hot stick and insulating rope.

Bare-hand Method

Methods have been developed to allow workers to work on conductors or other energized items as long as the platforms or buckets in which they are standing are insulated (see Fig. 2-44). Work may also be done from a dolly that rides on the conductor that is being worked on. In such an instance, no attempt is made to insulate the worker from the energized item. The worker now becomes energized together with the item. However, the support (or dolly) is insulated.

Extreme caution is necessary in this method. The insulated platform or bucket insulates the workers from a live conductor and ground, but will not protect them when working on two or more live conductors between which high voltages may exist. Although known as "bare-hand" method, workers usually wear noninsulating work gloves.

Review

• Structures for supporting overhead conductors are broadly classified as poles and towers. They may be made of natural wood: southern yellow pine, western red cedar, Douglas fir, larch, and other species. They may be chemically treated, round, tapered poles, or laminated, round or square poles (two or more layers of wood are glued together).

• Poles may also be of hollow, tapered, tubular design, made of steel or aluminum, or built up of flat metal members, latticed together, into a variety of cross sections. Reinforced concrete poles are used in special cases.

• Poles may be combined in A-frames, H-frames, and sometimes into V- or Y-type transmission structures.

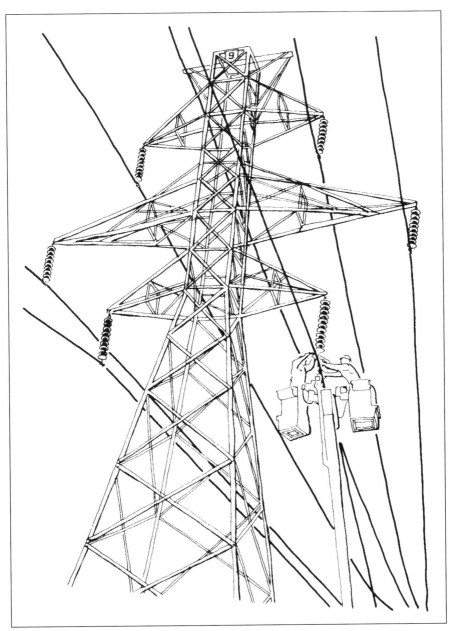

Fig. 2-44 Workers in insulated buckets repairing power lines

- Towers may be of several types:

  1. A tangent tower, on which the conductor supported is essentially a straight line
  2. A light-angle tower, on which the conductor supported changes direction slightly, perhaps 5° or 10°
  3. A medium-angle tower, with support changes of 20° to 30°
  4. A heavy-angle tower, which accommodates sharper turns

- Generally, guys and anchors should be installed on dead-ends, angles, long spans where pole strength may be exceeded, and at points of excessive unbalanced conductor tension.

- The choice of supporting structures for transmission lines is influenced by many factors, which, considered together, result in the greatest economy.

- Both skin effect and corona discharge may cause interference on communication circuits that may parallel the transmission lines. Corona discharge may also cause interference to local radio and television broadcasting. To lessen the effects of skin effect and corona discharge, hollow expanded conductors and corona rings are used in transmission lines.

- Cross-arms for towers are generally of galvanized steel or aluminum; wood is used for pole and A- or H-frame construction.

- In the ACSR type of cable, aluminum conductors are wrapped about a steel cable that provides strength.

- Insulators used for transmission lines are both of the suspension and pin or post types.

- Long overhead conductors of transmission lines are subject to aeolian vibrations and galloping (or dancing) conductors.

- Arcing rings and lightning arresters protect transmission lines from damage caused by lightning. Shielding the lines by means of a wire above them, which is directly connected to ground, is another form of protection from lightning. Lower tower footing resistance is important.

Study Questions

1. What are transmission line supports usually called? Of what material may they be made?
2. What type of transmission structures can be made of wood or metal poles?
3. Into what types may towers be classified?
4. Where are the guys and anchors generally installed?
5. What factors influence the choice of supporting structures?
6. What is meant by "skin effect"? What is corona discharge? How may these effects be minimized?
7. Of what materials are overhead transmission conductors made?
8. What types of insulators are used on transmission lines? What are the advantages of each?
9. How do the forces of nature affect overhead transmission lines?
10. How are overhead transmission lines protected from lightning?

3

Underground Construction

General Concepts

Underground transmission lines have some advantage of freedom from above ground weather and traffic problems, and thus experience fewer interruptions than overhead lines. However, there are a variety of failures that do affect underground cables. Interruptions underground may last from a few days to several weeks while the fault is located, the cable exposed, and repairs made under necessarily very exacting conditions.

This repair time compares unfavorably to that of outages on overhead lines, where fault location and repair are generally short-lived. Significant economic problems develop, therefore, because underground systems require that more facilities be available to attain the same level of reliability as that of the overhead systems. Underground transmission facilities are many times more costly than overhead ones. Consequently, they are feasible only in special locations, such as metropolitan centers, where towers on high poles in congested areas would be completely unacceptable.

Underground transmission cables may be laid directly in the ground in open fields or under suburban and rural roads (see Fig. 3–1). They may be drawn into ducts or installed in pipes in more or less densely populated urban and suburban areas (see Fig. 3–2). They can also be installed in troughs in special cases, such as under bridge structures or in tunnels (see

Fig. 3-1 Underground transmission cables buried directly in the ground

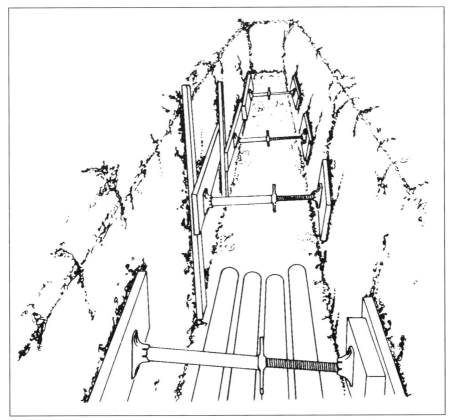

Fig. 3–2 Underground transmission cables installed in ducts

Fig. 3-3). The method of installation depends on the voltage and kind of cable, and the area in which it is installed.

Cables

Tracking

High-voltage cables present particular problems, since the conductors of the cables are subjected to thermal expansion and contraction from the loading and unloading of current during daily or other periodic cycles. They initiate motion with respect to the insulation, which tends to cause voids or pockets to form in the insulation. A void between the conductor and the insulation, or between the insulation and the grounded sheath, may also be

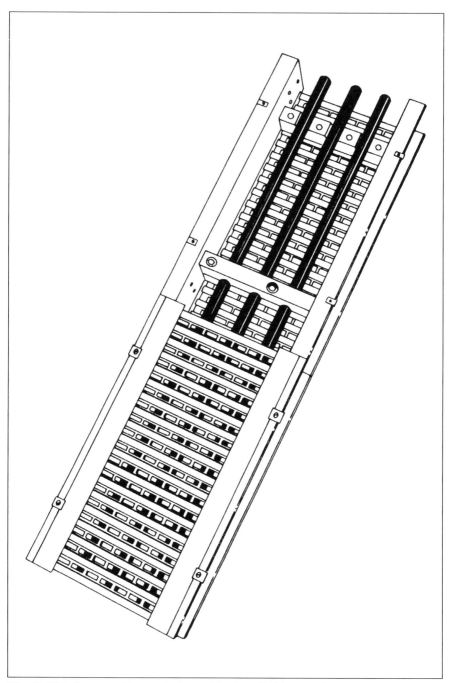

Fig. 3–3 Underground transmission cables installed in troughs

formed due to other factors. These include faulty manufacture, bending too sharply during installation, or thermal expansion and contraction because of load cycling (see Fig. 3-4a).

Such minute air pockets, under the electrostatic forces of the energized conductor, tend to ionize, that is, become conductors of electricity. This ionization of minute particles within the void causes corona discharge.

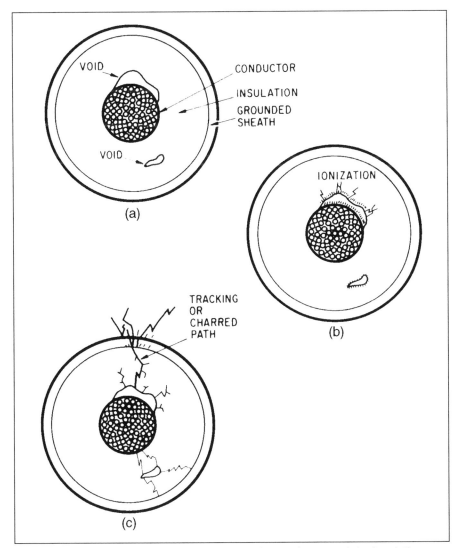

Fig. 3-4 Tracking in a cable: (**a**) void between the conductor and the insulation; (**b**) corona discharge; (**c**) cable failure

The corona discharge results in minute scorching and carbonizing of adjacent insulation. This is the beginning of "tracking" and the creation of ozone that damages the insulating value of most compounds (see Fig. 3-4b). When enough air pockets form, a tracking or charred path occurs where the insulation breaks down. Ultimately, the insulation is bridged with a carbonized track that is conductive, and the cable fails (see Fig. 3-4c). Some high-voltage cables for underground transmission systems are shown in Figure 3-5.

Insulating Materials

Insulation for cables has largely been oil-impregnated paper, though more recent cables employ plastic materials. Some cables consist merely of paper wrapped around one or more conductors, enclosed within a protective sheath. In others, the insulation wrapped around the conductors is subject to oil or gas under pressure. The gas used is usually nitrogen, but sulfur hexafluoride (SF_6) has also been used. Theoretically, when such voids occur in the insulation, the oil or gas under pressure fills the void and prevents ionization, tracking, and failure. Figure 3-6 shows the most suitable voltage ranges for paper-insulated cables.

Historically, fluid-impregnated paper-tape insulation has been the workhorse of high-voltage transmission cables. Its excellent properties result

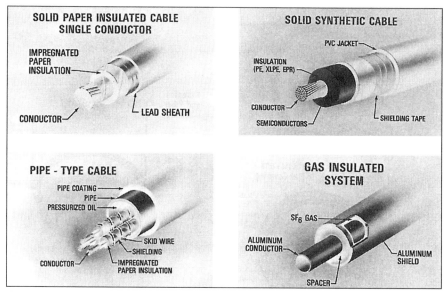

Fig. 3-5 Some high-voltage cables for underground transmission systems

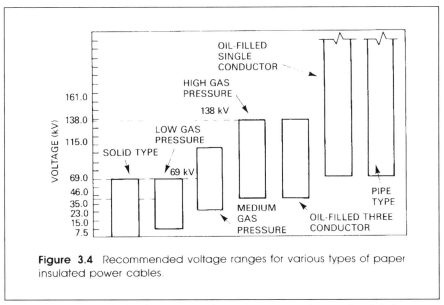

Figure 3.4 Recommended voltage ranges for various types of paper insulated power cables.

Fig. 3-6 Recommended voltage ranges for various types of paper-insulated power cables

in very reliable underground transmission operation. In the mid-1980s, a new, laminated polypropylene and paper-tape (PPP) insulation was made commercially available. It featured properties that improved on those of paper insulation—thinner insulation walls, lower dielectric loss, and higher impulse level strength. PPP-insulated cables have higher capacity ratings than all paper cables. For example, a 1.5 thousand circular mills (mcm) copper conductor paper cable would be rated at 490 MVA, compared to PPP rated at 570 MVA, which is about 15% more.

Hollow Cables

In "hollow" cables, the oil or gas passage is located within the conductor or cable (see Fig. 3–7). In "pipe" cables, the oil or gas surrounds the conductors (see Fig. 3-8). Special accessories and auxiliary equipment to handle the oil or gas are needed with these latter cables. A comparison of power transmission capacity between gas- and oil-impregnated insulated cables is shown in Figure 3-9.

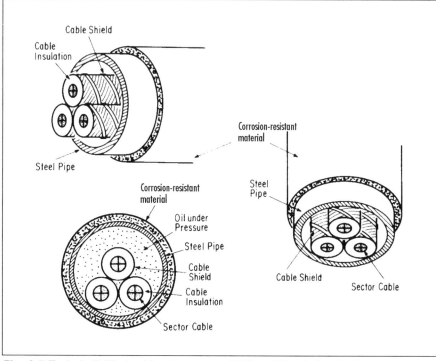

Fig. 3-7 Typical oil-filled cables (*courtesy of Edison Electric Institute*)

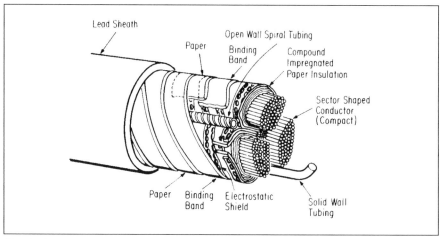

Fig. 3–8 Typical gas-filled cables (*courtesy of Edison Electric Institute*)

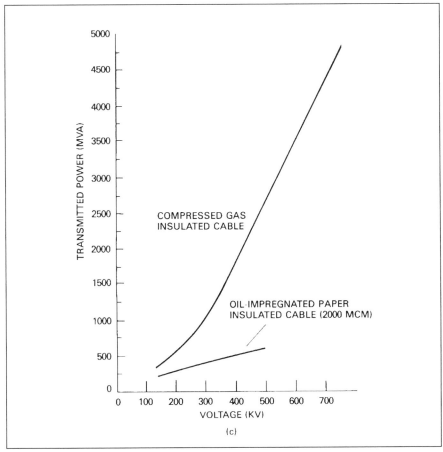

Fig. 3–9 Comparison of power transmission capability of compressed gas, insulated cable, and oil-impregnated paper insulated cable (*courtesy of Edison Electric Institute*)

Solid Insulation Cables

Solid insulated cables were formerly restricted to voltages of about 46 kV. However, with improved methods of manufacture and use of plastics, such as polyethylene-polypropylene, solid cables of 69 kV and 138 kV have been built and put into operation.

Pipe Coverings

Hollow cables have been used in circuits operating from 69 kV to 350 kV. These installations require extraordinary care, since the pipe interior must be free of contamination. Evacuation before the introduction of the oil or gas therefore requires a high degree of vacuum and cleanliness. Special coverings on the pipe are used to avoid corrosion, electrolysis, and other damage to the pipe. Obviously such installations, while varying with type of construction, the nature of the ground, road surface, transportation, and other facilities, are extremely expensive.

Cable Installation

Care should be exercised when installing cables by pulling them in a duct or pipe. This is especially necessary in the case of gas- and oil-type cables. The stresses set up in the cables during the operation may cause damage to the conductors, the insulation, and the gas or oil paths. Outer armor wires, usually spiraled about a cable, may tighten about a cable so as to damage the insulation and cut off the flow of gas or oil in the cable. Thousands of feet of gas and oil cables may be pulled in at one time. Dynamometers, instruments for measuring tension, are often employed during the pulling process to ascertain that allowable pulling stresses are not exceeded (see Fig. 3-10).

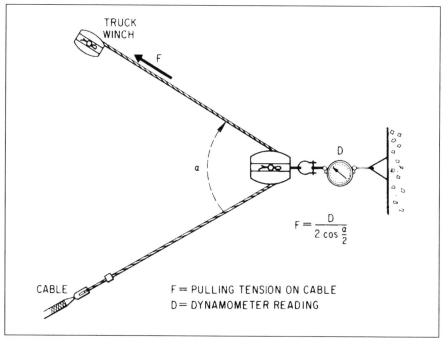

$$F = \frac{D}{2 \cos \frac{a}{2}}$$

F = PULLING TENSION ON CABLE
D = DYNAMOMETER READING

Fig. 3-10 A dynamometer

Repair of Oil-filled Cables

While solid cables require relatively little maintenance and are comparatively easy to replace or repair, oil- and gas-filled cables present much greater difficulties. Oil or gas leaks occasionally occur, and these must be found and repaired, generally without deenergizing the cable. At some point, oil-filled cables have to be replaced, whether because of failure, inadequacy, public improvement, or for other reasons. When this occurs, it is necessary to seal off the flow of oil on both sides of the section on which work is to be performed.

The flow of oil is sealed off by freezing an oil slug from the outside of the cable or pipe by pouring liquid nitrogen around it until there is assurance the oil flow is stopped. These low temperatures are maintained by encasing the cable or pipe in ice and continued dripping of liquid nitrogen. The cable or pipe may then be cut, cables replaced, repaired, or rerouted, and in the case of pipe cables, the pipe assembly welded together again.

Care is taken to reestablish the protective covering. Before reenergization, it is necessary not only to replace the oil in the affected section, but to ascertain that all of the oil in the system is free from air, moisture, and other contaminants. This is a time-consuming and expensive process.

Figure 3-11 illustrates how freezing is employed. The method shown consists of an approximate 2-ft-long split lead sleeve with two stand pipes. The sleeve is opened sufficiently to pass round the cable. The gap is then closed and sealed along its length with solder. The ends are then sealed to

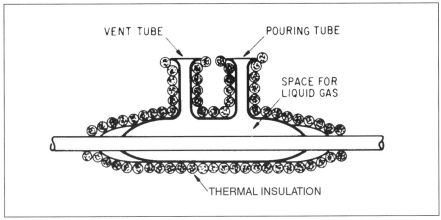

Fig. 3-11 Freezing a section of an oil-filled cable to isolate a leak

the cable jacket by the application of rubber tapes and reinforcing tapes (such as hessian). The whole of the sleeve and the stand pipes are then insulated thermally either in tape or rope form.

Liquid nitrogen is poured into the freezing sleeve, via a funnel, to one of the stand pipes; the other acts as a vent for the exhaust of the nitrogen vapor. Experience has shown that the time required to provide a solid blockage in a cable of approximately 2-1/2 in. outer diameter is 30 minutes. For cables up to approximately 4 in. in diameter, 1 hour is needed.

Repair of Gas-filled Cables

Repair of gas-filled cables generally follows the same procedure as for oil-filled cables except for one important difference. There is no oil to be frozen to stop the flow of oil out of the cable while it or a joint is being repaired. In gas cables, the gas is allowed to escape freely from between two stop joints and replaced with clean, dry gas when work is completed. Gas is fed at one of the stop joints and allowed to flow out at the stop joint on the other side of the repair. Gas is allowed to flow out for a brief period to ascertain no air or moisture remains in the cable before sealing at both stop joints.

Joints

In long installations, it is obvious that a gas or oil leak in either the hollow or pipe cable systems could lead to a long and costly decontamination process. To restrict the possibility of widespread contamination, such systems are usually sectionalized through the insertion of stop joints and semistop joints in the cable system. The number and types of such joints are usually dependent on the length of the line and its importance. The stop joint provides a physical barrier to both the conductor and the gas or oil flow system. These are used on long transmission lines and permit the line to be sectionalized so that the cables may be repaired or replaced without affecting the entire length of line.

The semistop joint allows the conductor to pass through but imposes a physical barrier to the gas or oil-flow system (see Fig. 3-12). Hence, in the event of a leak that would allow air or other contaminant to enter the cable, only a relatively small section of the line between the semistop joints is affected. In repairing gas-filled cables, the gas in one of the sections is completely replaced, and no operation similar to the oil freezing is required.

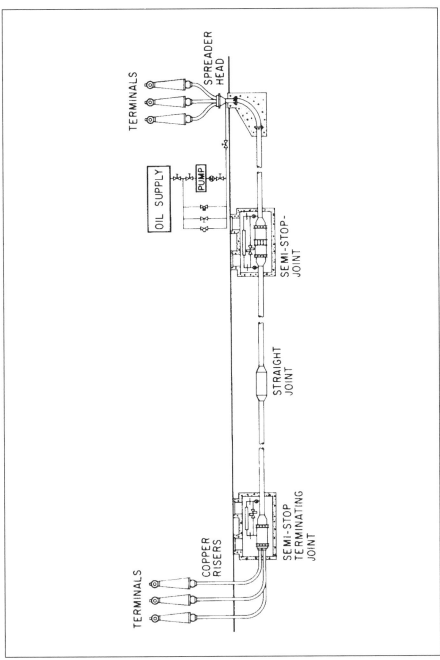

Fig. 3-12 Isolating a cable section with a semistop joint

Hence, both the time for repair and decontamination and the cost for such an incident are held down.

Direct Burial

When lead-sheathed or steel-armored cables are laid directly in the ground, precautions should be taken to prevent their being damaged. Stones and other rough material should be kept away from the cable. It is important to dig the trench deeply enough to keep plows or diggers from damaging the cable. It also should be wide enough to enable soft, sandy fill to be packed around the cable to prevent bruises and cuts. This will also improve heat dissipation, described later in this chapter. A plank or other marker may be placed horizontally in the trench on top of the cable to act as a warning to workers. It will also serve to protect the cable in the event the ground is disturbed after the cable is laid (see Fig. 3-13).

Manholes

Manholes may or may not be provided, depending on the area of installation, length of cable runs, and other factors. They are usually required for pulling-in purposes and for splices both for original installation and for repair or replacement. Their dimensions and arrangement should be such that the cables are not bent too sharply. However, they must allow for cable

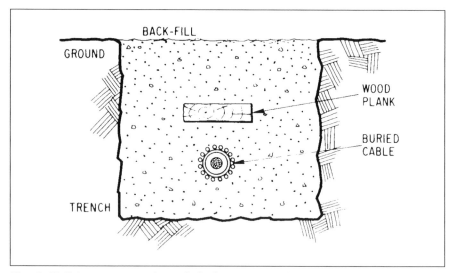

Fig. 3-13 Using a creosoted wood plank to protect a buried cable

movement under load or temperature differences so as not to put undue stress on the splice. Typically, for a 3-in. diameter cable, as shown in Figure 3-14, offset length (L) would be 66 in., joint length, 30 in., and the manhole length, 162 in. In many instances, manholes have been eliminated, and the entire installation is buried in the ground.

Soils

Underground cables—whether directly buried (see Fig. 3-13) or encased in a steel pipe (see Fig. 3-15)—are directly affected in capacity rating by the thermal characteristic of the surrounding soil. Thermal resistivity is the measure of the ability of the soil to conduct heat away from the cable source and be dissipated. Even soils with poor resistivity characteristics, such as dry clay, can achieve acceptable levels of resistivity if the moisture content is high enough. Moisture in the earth tends to move away from the cable heat source to an ambient point some distance from the cable.

If the moisture does not migrate back towards the cable fast enough, the earth surrounding the cable will dry out and become a poor thermal conductor. If the drying out continues, a hot spot can develop through "thermal runaway," with the probability of cable failure. This can be prevented if the load on the cable is reduced or the thermal properties of the soils are improved by proper application of a better backfill, such as thermal sand.

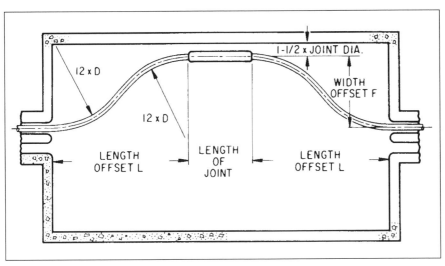

Fig. 3-14 A typical manhole

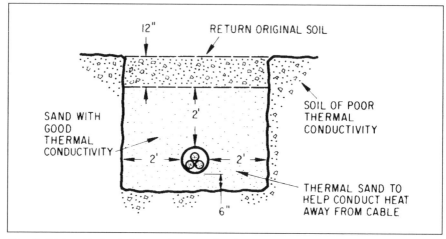

Fig. 3-15 A cable buried in underground soil

Splices

In splicing such cables, precautions similar to those used with over-head splices are taken. These include filling in the indentations on compression-type connectors and smoothing the surfaces to reduce or eliminate corona effects. The conductors are usually of stranded copper, though they may be of aluminum. Hollow conductors lend themselves to reducing skin effect, similar to overhead conductors.

Charging Currents

In underground transmission systems, the high voltage and the configuration of the conductor and metallic sheath produce a capacitor effect, a phenomenon somewhat similar to the condition of clouds in thunderstorms that produce lightning. This is intensified as voltage levels increase, resulting in so-called charging currents that may exceed the cable thermal limits. This occurs if the length of the cable is sufficiently great, which makes the amount of current necessary to "charge" the cable rather large, thereby reducing its load-carrying capability. In instances where voltage levels are in the order of 345 kV or greater, total depletion of the load-carrying ability may occur after only 30 miles of transmission.

Corrective equipment, called a shunt reactor, is available to compensate for this energy loss, but at extremely high cost. This cost must be added to the high cost of underground cable, the cost of splicing (which requires special skills), and other costs associated with underground installations. Figure 3-16 illustrates the charging current phenomenon. For a more technical explanation, refer to appendix C, "Basic Electricity."

Review

• Underground transmission cables may be: (1) placed directly in the ground using armored cables; (2) drawn into ducts or installed in pipes; or (3) installed in troughs in special cases. Insulation for cables has largely been oil-impregnated paper, though more recent cables employ plastic materials.

• In "hollow" cables, the oil or gas passage is located within the conductor or cable; in "pipe" cables, the oil or gas surrounds the conductors.

• Dynamometers are often employed while pulling cables to assure that allowable pulling stresses are not exceeded.

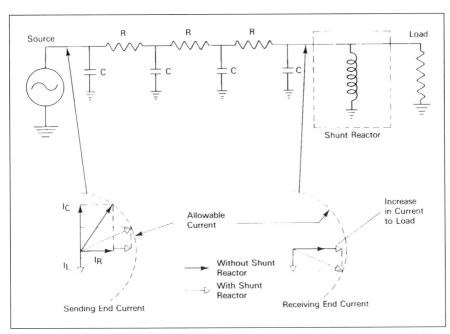

Fig. 3-16 Simplified cable circuit shows how power delivered to load is affected by a shunt reactor; in the absence of shunt reactors, this power decreases to zero at some critical length

• Solid cables require relatively little maintenance and are comparatively easy to replace or repair, while oil- and gas-filled cables present much greater difficulties. Where oil-filled cables have to be replaced, it is necessary to seal off the flow of oil on both sides of the section on which work is to be performed.

• Stop joints are used in long installations to sectionalize the cable system to facilitate repair.

• Manholes are often required for pulling-in purposes and for splices both for original installation and for repair or replacement.

• Underground cables, whether of the types buried directly in the ground or pipe types, are installed at depths below the frost line, usually 24 in. deep or deeper. The soils in which they are buried must provide adequate heat dissipation to prevent overheating and failure of the cables. If hot spots occur, that particular area may be replaced with earth or "thermal" sands having better heat-dissipating characteristics.

• In splicing underground cables, precautions of filling in the indentations on compression connectors and smoothing the surfaces to reduce or eliminate corona effects must be taken.

• In underground transmission systems, the high voltage and the configuration of the conductor and sheath produce a condenser action. This is intensified as voltage levels increase, resulting in "charging currents" that may exceed the cable thermal limits.

Study Questions
1. How may underground transmission cables be installed?
2. What do underground transmission cables consist of?
3. What may happen to insulation in underground cables, especially at higher voltages?
4. What may the insulation of underground transmission cables consist of?

5. What is the difference between a "hollow" cable and a "pipe" cable?
6. What are stop joints and semistop joints? Where are they used?
7. How are sections of oil-filled cables replaced or repaired?
8. What is the function of manholes? What determines their design?
9. What effect does the surrounding soil have on buried cables?
10. What precaution should be taken when splicing two conductors?

Substations

Substation Functions and Types

Substations serve as sources of an energy supply for the local areas of distribution in which they are located. Their main functions are to receive energy transmitted at high voltage from the generating stations, reduce the voltage to a value appropriate for local use, and to provide facilities for switching (see Fig. 4-1).

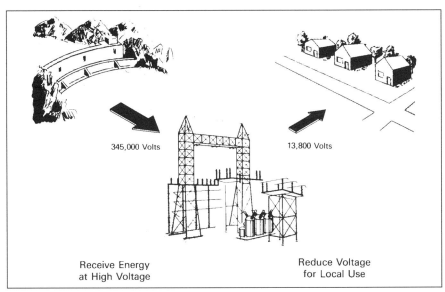

345,000 Volts   13,800 Volts

Receive Energy
at High Voltage

Reduce Voltage
for Local Use

Fig. 4–1 Some functions of a substation

Substations have some additional functions. They provide points where safety devices may be installed to disconnect circuits or equipment in the event of trouble. Voltage on the outgoing distribution feeders can be regulated at a substation. In addition, a substation is a convenient place to make measurements to check the operation of various parts of the system.

There are substations that are simply switching stations, where different connections can be made between various transmission lines.

Some substations are entirely enclosed in buildings; others are built entirely in the open (see Fig. 4-2). In this latter type, the equipment is usually enclosed by a fence. Other substations have step-down transformers,

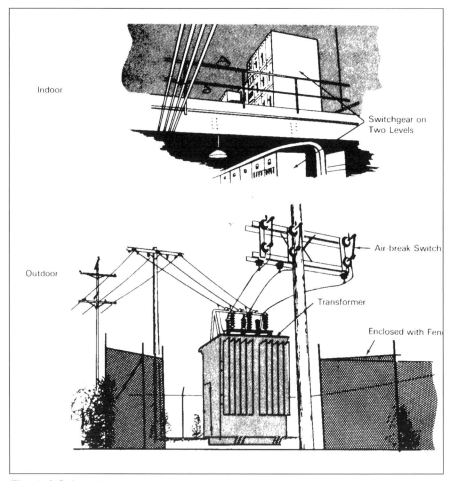

Indoor

Switchgear on
Two Levels

Air-break Switch

Outdoor

Transformer

Enclosed with Fen

Fig. 4–2 Substation locations

high-voltage switches, circuit breakers and lightning arresters located just outside the substation building. The substation building houses the relaying and metering facilities.

Factors Influencing Location

Sites for distribution substations are generally selected so that the stations will be as near as possible to the load center of the areas they are intended to serve. Sites for transmission substations are selected using the following criteria:

- Terminal points of interconnections
- Transitions between overhead and underground transmission
- Key points of transmission interfaces for transfer of power between areas
- Critical service reliability locations of the transmission system

Typical Features of a Transmission Substation

Transmission substations usually have several incoming and outgoing transmission lines connected to bus arrangements through very high-speed circuit breakers. Bulk power distribution in various directions is the principal purpose but may be accompanied by distribution stepdown transformers and circuit breakers to serve local loads. Stepdown of a high-transmission voltage to a subtransmission system may also be present.

Usually these stations are operated automatically, with control communication routed back to an operating center. Such centers not only tell operators the condition of the stations, but enable them to operate equipment remotely. These control communication circuits may be utility-owned wires, public telephone circuits, or radio or microwave links.

The design of the substation arrangement should permit taking lines or equipment out of service for maintenance or operating purposes without affecting service continuity. A typical substation arrangement is shown in Figure 4-3, using standard symbols for equipment.

Automatic Versus Manual Operation

Substations may have an operator in attendance part or all of the day, or they may be entirely unattended. In unattended substations, all equipment functions automatically, or may be operated by remote control from an

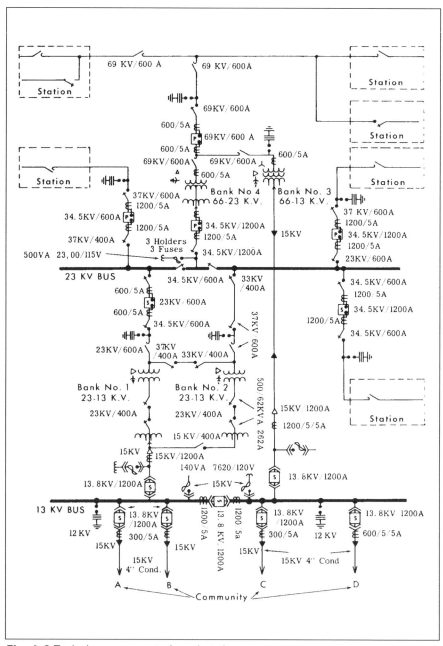

Fig. 4–3 Typical arrangement of a substation

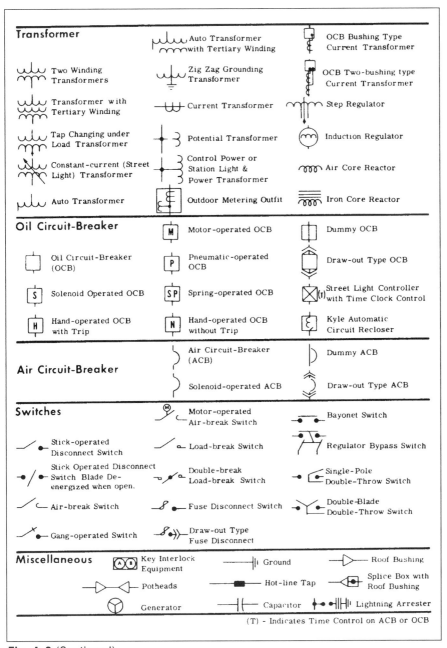

Fig. 4–3 (Continued)

attended substation, or from a control center (see Fig. 4-4). In some unattended substations, the equipment function is monitored and will give an alarm at a remote station when a "roving operator" needs to be dispatched to operate that station.

Substation Equipment—The Transformer

When a transmission substation is a multipurpose station, the voltage of the incoming supply is changed to that of the subtransmission or distribution feeders by means of a transformer.

Fundamentally, a transformer consists of two or more windings placed on a common iron core (see Fig. 4-5). All transformers have a primary winding and one or more secondary windings. The core of a transformer is made of laminated iron and magnetically links the coils of insulated wire that are

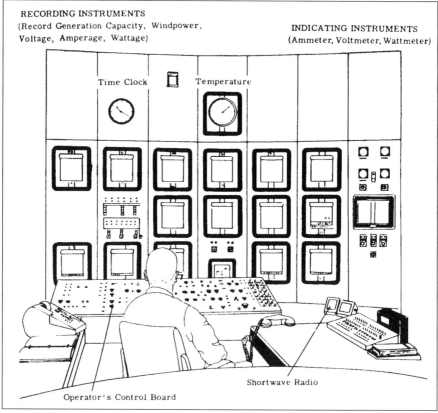

RECORDING INSTRUMENTS
(Record Generation Capacity, Windpower, Voltage, Amperage, Wattage)

INDICATING INSTRUMENTS
(Ammeter, Voltmeter, Wattmeter)

Time Clock   Temperature

Shortwave Radio

Operator's Control Board

Fig. 4–4 A control center that operates several substations by remote control

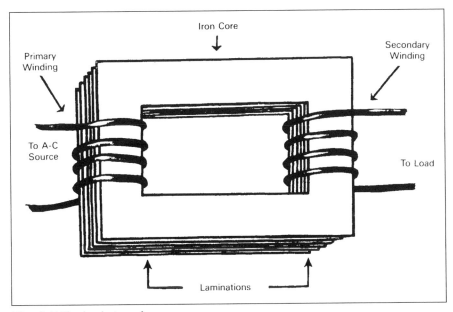

Fig. 4–5 The basic transformer

wound around it. There is no electrical connection between the primary and the secondary windings; the coupling between them is through magnetic fields. This is why transformers are sometimes used for no other purpose than to isolate one circuit from another electrically. When this is done, the transformer used for this purpose is called an "isolating" transformer.

The winding that is connected to the source of power is called the primary, and the winding connected to the load is called the secondary. The essential function of the conventional power transformer is to transfer power from the primary to the secondary with a minimum of losses. As we shall see, in the process of transferring energy from primary to secondary, the voltage delivered to the load may be made higher or lower than the primary voltage.

Step-up and Step-down Transformers

Transformers are wound to be as close to 100% efficient as possible. That is, all the power in the primary (as much as possible) is transferred to the secondary. This is done by selecting the proper core material, winding the primary and secondary close to each other, and following a number of other careful design specifications.

Thus, assuming 100% transformer efficiency, we can then assume that the relationship between the primary and secondary voltages will be the same as the relationship between their turns. If the secondary has more turns than the primary, we say that the transformer is operating as step-up transformer. If the secondary has less turns than the primary, we say that the transformer is operating as a step-down transformer.

Figure 4-6 is a representation of an autotransformer together with a conventional transformer. The autotransformer is somewhat different from the conventional transformer in that a portion of the primary and secondary is common, or makes use of the same turns. However, like all other transformers, the autotransformer does have the basic primary and secondary windings, although not physically isolated from each other.

Turns Ratio

Whether a transformer is of a step-up or step-down type, the power in the primary is equal to the power in the secondary (see Fig. 4-7). Thus, if the load draws 1,000 watts, the product of voltage and current in the primary is also equal to 1,000 watts. Another important principle is the fact that the primary and secondary voltages are in the same ratio as their turns. If the

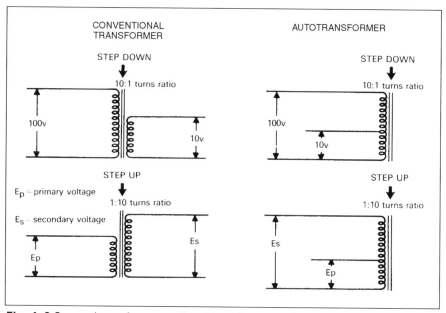

Fig. 4–6 Comparison of a conventional transformer and an autotransformer

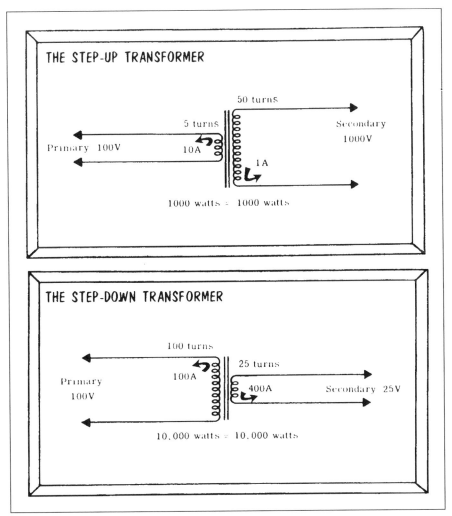

Fig. 4–7 Step-up and step-down transformers

secondary has twice the turns of the primary, the secondary voltage will be twice as great as that of the primary.

Transformer Rating

The nameplate on a transformer gives all the pertinent information required for the proper operation and maintenance of the unit (see Fig. 4-8). The capacity of a transformer (or any other piece of electrical equipment) is limited by the permissible temperature rise during operation. The heat gen-

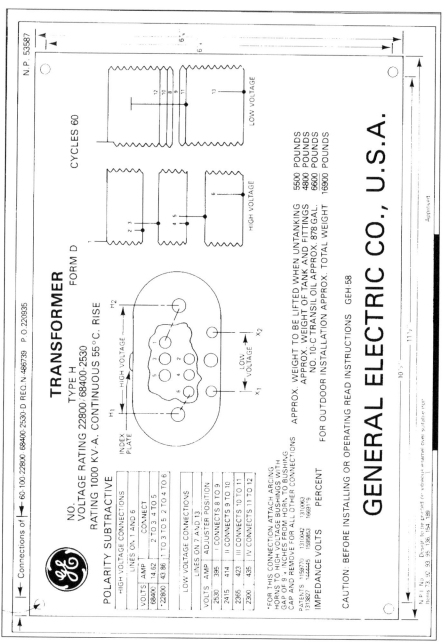

Fig. 4–8 Typical transformer nameplate (*courtesy General Electric Company*)

erated in a transformer is determined by both the current and the voltage. Of more importance is the kilovolt-ampere (kVA) rating of the transformer. This indicates the maximum power on which the transformer is designed to operate under normal conditions. Other information generally given on the nameplate includes:

- The phase (single-phase, three-phase, etc.)
- The primary and secondary voltages
- The frequency
- The permitted temperature increase
- The cooling requirements (includes the number of gallons of fluid that the cooling tank may hold)
- Percent impedance (full load voltage drop)

Primary and secondary currents may be stated at full load.

Depending upon the type of transformer and its special applications, there may be other types of identifications for various gauges, temperature indication, pressure, drains, and various valves.

The transformer consists primarily of a primary and a secondary winding; however, there are many other points to take into consideration when selecting a transformer for a particular use.

Methods of Transformer Cooling

The wasted energy in the form of heat generated in transformers because of unpreventable iron and copper losses must be carried away. This is necessary to prevent excessive rise of temperature and injury to the insulation around the conductors. The cooling method used must be capable of maintaining a sufficiently low average temperature. It must also be capable of preventing an excessive rise in any portion of the transformer, and the formation of hot spots. This is accomplished, for example, by submerging the core and coils of the transformer in oil, and allowing free circulation for the oil (see Fig. 4–9).

Since oil may be a fire hazard, inert fluids, known as askarels, are sometimes used in place of the oil. These special fluids may be harmful to personnel handling them as well as to the varnishes generally applied to the

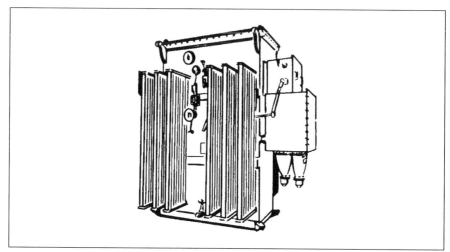

Fig. 4–9 Oil-cooled transformer

insulation of the coils. Extreme care should be exercised in handling to prevent contact with eyes or open cuts and wounds.

Some of the oils and askarels in use have been found to contain polychlorinated biphenyl (PCB), considered a teratogen. Steps have been taken to eliminate the hazard by replacing these oils and askarels with safer oils and other coolants. It also requires disposal of these materials as "regulated hazardous wastes" under the federal Toxic Substances Control Act (TSCA). In some instances this is accomplished by replacing existing transformers. In other instances, the contaminating fluid may be drained at the site. The transformer is flushed several times with special contaminant-absorbing fluids that draw the PCB from the transformer core and parts. The drained fluids are then replaced with PCB-free oil or other fluids.

In clean, dry, indoor locations, an open, dry-type air-cooled unit can be used. For outdoor (and indoor) use, a sealed, dry unit can be employed (see Fig. 4-10).

Some transformers (fluid filled or dry) are cooled by other means. These include forced air or air blast, a combination of forced oil and forced air, and in some special applications, water cooling.

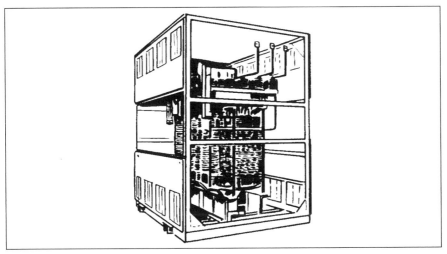

Fig. 4–10 Air-cooled transformer

Related Substation Equipment

Busbars

Busbar (or bus, for short) is a term used for a main bar or conductor carrying an electric current to which many connections may be made.

Buses are merely convenient means of connecting switches and other equipment into various arrangements. The usual arrangement of connections in most substations permits working on almost any piece of equipment without interruption to incoming or outgoing feeders.

Some of the arrangements provide two buses to which the incoming or outgoing feeders and the principal equipment may be connected. One bus is usually called the "main" bus and the other the "auxiliary" or "transfer" bus. The main bus may have a more elaborate system of instruments, relays, etc. associated with it. The switches that permit feeders or equipment to be connected to one bus or the other are usually called "selector" or "transfer" switches. As shown in Figure 4-11, busbars comes in a variety of sizes and shapes.

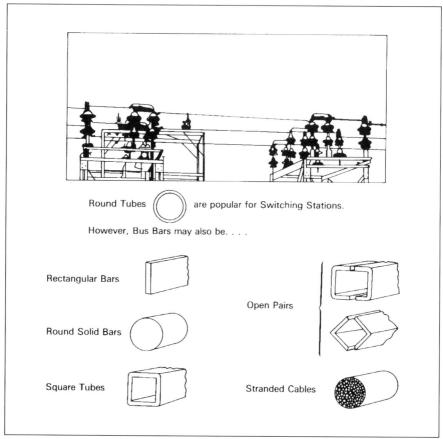

Round Tubes ◯ are popular for Switching Stations.
However, Bus Bars may also be. . . .

Rectangular Bars

Open Pairs

Round Solid Bars

Square Tubes

Stranded Cables

Fig. 4–11 Typical busbars

Regulators

A regulator is really a transformer with a variable ratio (see Fig. 4-12). When the outgoing voltage becomes too high or too low for any reason, the apparatus automatically adjusts the ratio of transformation to bring the voltage back to the predetermined value. The adjustment in ratio is accomplished by tapping the windings, varying the ratio by connecting to the several taps. The unit is filled with oil and is cooled much in the same manner as a transformer. A panel mounted in front of the regulator contains the relays and the other equipment that control the operation of the regulator.

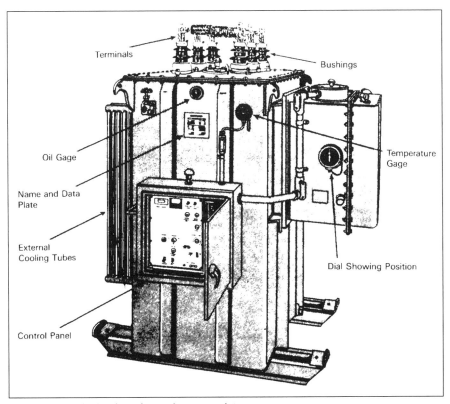

Terminals

Bushings

Oil Gage

Temperature Gage

Name and Data Plate

External Cooling Tubes

Dial Showing Position

Control Panel

Fig. 4–12 Typical substation voltage regulator

Circuit Breakers

In past years, oil circuit breakers (see Fig. 4-13) were used to interrupt load or "fault" current by making or breaking electrical contact. The oil served to quench the arc as the contacts opened. Transmission breakers are designed to operate very fast as compared to distribution circuit breakers. Usually, transmission circuit breakers operate automatically through relay protection arrangements, or manually by remote control from a nearby control panel in the substation. They can also be operated from a remote location by communication with a power control center.

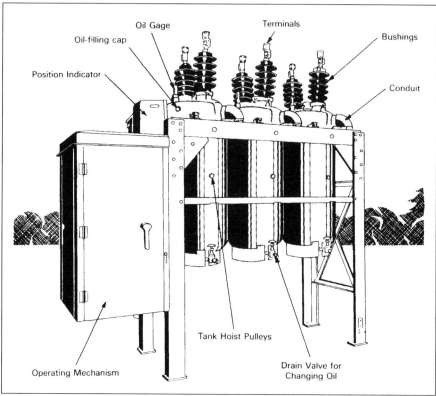

Oil Gage

Oil-filling cap

Terminals

Bushings

Position Indicator

Conduit

Operating Mechanism

Tank Hoist Pulleys

Drain Valve for
Changing Oil

Fig. 4–13 Typical oil circuit breaker

Air Break and Disconnect Switches

Some switches are mounted on an outdoor steel structure called a rack (see Fig. 4-14 and Fig. 4–15), while others may be mounted indoors on the switchboard panels. They are usually installed on both sides of a piece of equipment to deenergize it effectively for maintenance.

Instrument Transformers

When values of current or voltage are large, or when it is desired to insulate the meter or relay from the circuit in which they are to operate, an instrument transformer is used.

In measuring current of high value, a current transformer (CT) is used (see Fig. 4-16). The ratio of transformation reduces the high-current circuit, which in this case is the primary of the transformer, to a smaller current in the secondary connected to the ammeter or relay.

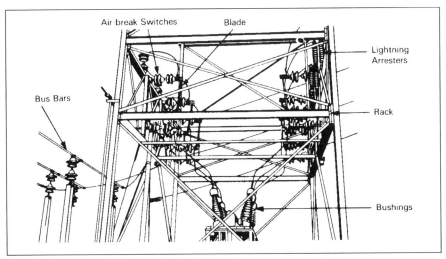

Fig. 4–14 Air-break switches mounted on a substation rack

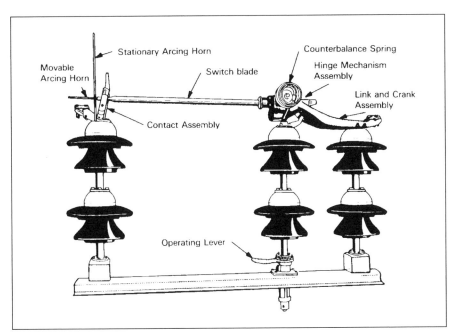

Fig. 4–15 Air-break switch assembly

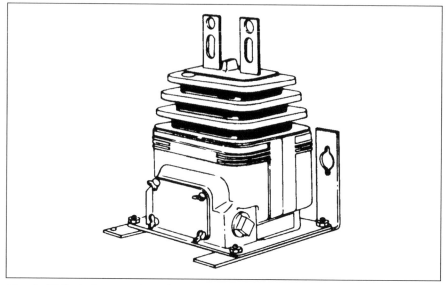

Fig. 4–16 Typical current transformer

Because of the high turn ratio of the current transformer, a very high (and very unsafe) voltage may exist at the secondary terminals with the high voltage side energized and the secondary open at the terminals. As soon as a load (instruments or relays) is connected to the terminals the voltage drops very rapidly. For safety reasons the terminals should always be short-circuited when no loads are connected. Loads are first connected before the short-circuiting device is removed.

Similarly, a potential transformer (PT) (see Fig. 4-17) has a fixed ratio of primary to secondary voltage. The secondary terminals are connected to the voltmeter or relay circuit.

Instrument transformers differ from power transformers in that they are of small capacity and are designed to maintain a higher degree of accuracy under varying load conditions.

Relays

A relay is a low-powered device used to activate a high-powered device. In a transmission or distribution system, it is the job of relays to give the tripping commands to the right circuit breakers.

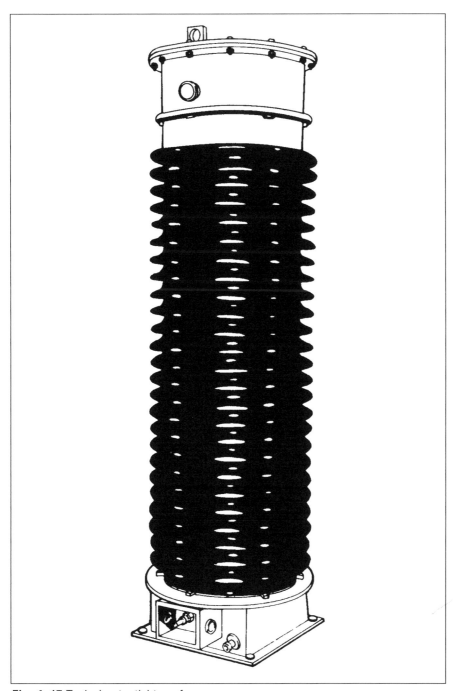

Fig. 4–17 Typical potential transformer

The protection of the lines and equipment is of paramount importance and is usually accomplished by the opening of circuit breakers automatically actuated by relays. In general, it is more important to provide protection for the components of a transmission system than on a distribution system. Greater blocks of load may be affected and resultant damage to lines and equipment may be more costly.

Relays are used to protect the feeders and the equipment from damage in the event of fault. In effect, these relays are measuring instruments, but equipped with auxiliary contacts that operate when the quantities flowing through them exceed or go below some predetermined value (see Fig. 4-18). When these contacts operate, they in turn actuate mechanisms that usually operate switches or circuit breakers, or in the case of the regulator, operate the motor to restore voltage to the desired level.

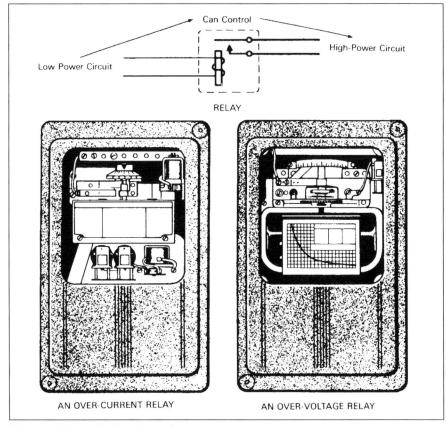

Fig. 4–18 Typical substation relays

With the solid-state electronic technology currently available and associated computer development, the relay-protection system has become more flexible, reliable, and adaptive to changing conditions than was possible with electromechanical relays. System Control and Data Acquisition (SCADA) systems, coupled with improved relay technology and computer information processing, are valuable tools to utilities. They provide system operators with the ability to control and coordinate the complex power systems while adjusting to competition transactions, open access to transmission, and the growth of independent power producers.

Bus Protection Differential Relaying

In providing protection against faults or buses, current supplied to the bus is measured against current flowing from it (see Fig. 4-19). These should be equal (with slight tolerances). When a fault develops on such a bus, this balance is disturbed, and a relay will operate, usually clearing both incoming and outgoing feeds from the bus. This is known as differential relaying.

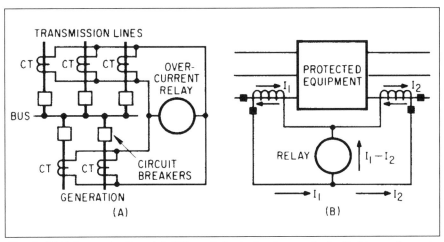

Fig. 4–19 Relay protection: (**a**) Differential relaying is usually limited to equipment concentrated in one area due to the large numbers of control wires between sensing devices or current transformers; (**b**) A difference in current on either side of the protected equipment, due to a fault in the protected equipment, will be detected by current transformers and will pass through the relay actuating coil ($I_1 - I_2$). When $I_1 = I_2$, no current passes through the relay.

Transmission Line Protection

Relay Protection

When a fault occurs on a transmission line, from whatever cause, it is imperative that the line be deenergized quickly for safety reasons. This is accomplished by the opening of circuit breakers at each of its terminals, which are actuated by relays. The relays contain a movable element that is actuated by the current flowing through it. The movable element makes or breaks one or more sets of contacts. These, in turn, activate the mechanisms that operate the circuit breakers.

The minimum value of the current that actuates the relays is predetermined for the basic overcurrent relay. When an additional element is added that will actuate the relay when the direction of current flow is opposite to the normal flow, the relay is termed a directional relay. When the movable element is actuated by a difference in the value of the current flowing into the protected line or equipment and that away from it, the relay is called a differential relay. Refinements in the applications of the currents flowing in the relay result in relays sensitive to other features of line and equipment operation. These are discussed more fully in a companion book, *Electrical Transformers and Power Equipment* (Pansini, 1988).

Applications of such protective relays are shown in Figure 4-20 for typical lines emanating from a generating station supplying one or more substations.

When a fault occurs on a transmission line near one of its terminals, the greater part of the fault current will flow through the circuit breaker of the nearest terminal. This will cause the circuit breaker to operate first, making the other end supply the fault current until the circuit breaker farthest from the fault operates somewhat later. The relatively much longer period in which fault current flows may cause severe damage. Hence, it is desirable that the circuit breakers at both ends of a faulted transmission line be opened simultaneously. This is especially true in the case of rather long transmission lines.

Pilot Protection

The simultaneous opening of circuit breakers at the terminals of transmission lines may be accomplished by means of a communication link between the circuit breakers involved.

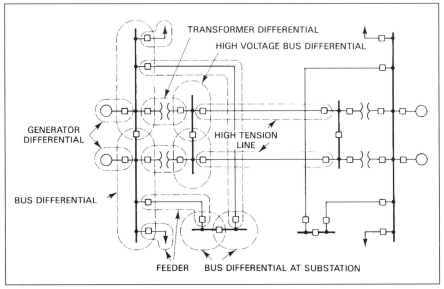

Fig. 4–20 Protective zones for generators and outgoing high-voltage transmission lines (*courtesy Westinghouse Electric Company*)

This link may consist of physically separate pilot wires. Two schemes employing overcurrent relays at each end are shown in Figure 4-21a and Figure 4–21b. A scheme employing differential relays at one end is shown in Figure 4-21c. All of these employ three to six pilot wires. One scheme employing only two pilot wires, polyphase directional relays at each end, and a direct current source is shown in Figure 4-21d.

Another more popular method requiring only two pilot wires employs an AC source and special relays. These combine the currents in each of the current transformers into a single-phase voltage that is compared to a similar quantity from the opposite end of the line. A simplified circuit is shown in Figure 4-22.

Often pilot wires are leased, adding to the cost of such systems. Moreover, for positive and effective operation, the systems have a practical limit, usually about 10 miles. For longer lines, carrier pilot relaying and microwave relaying systems are employed.

Carrier Pilot Relaying

In this type of protection, the pilot wires are replaced by the transmission line conductors themselves, with a high-frequency current (50–200

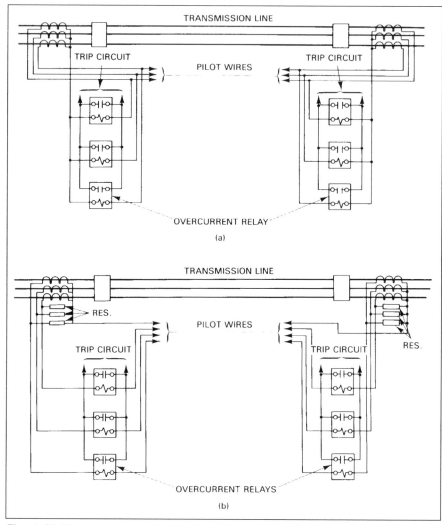

Fig. 4–21 Pilot wire schematic: (**a**) circulating-current pilot wire schematic with load currents and through fault currents circulating over pilot wires; (**b**) balanced-voltage pilot-wire schematic with load currents and through fault currents, producing equal opposing voltages at line terminals; (*courtesy Westinghouse Electric Company*)

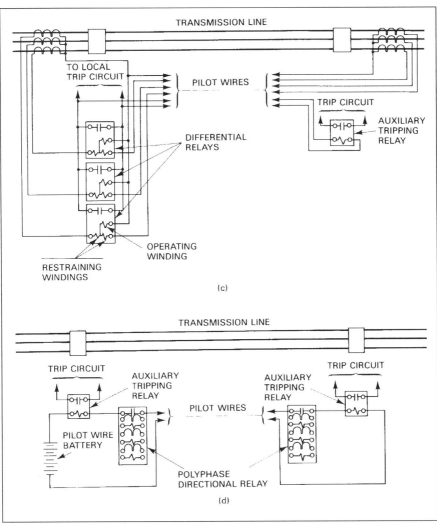

Fig. 4–21 Pilot wire schematic: (**c**) percentage differential relays pilot-wire schematic; (**d**) directional comparison pilot wire schematic using DC over a pair of wires (connections a–c omitted for simplicity)(*courtesy Westinghouse Electric Company*)

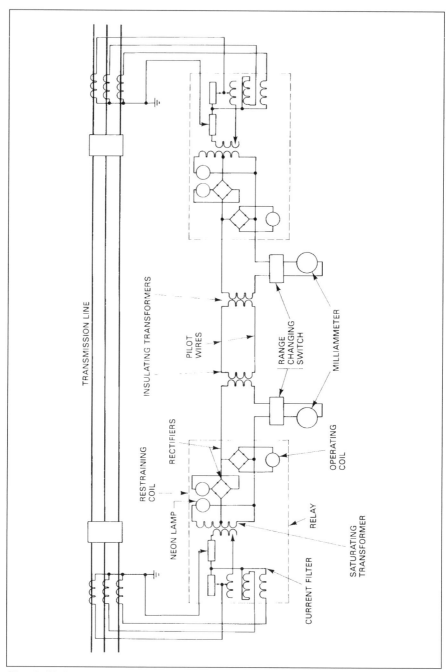

Fig. 4–22 Alternating current pilot wire schematic using special relays (*courtesy Westinghouse Electric Company*)

kilocycles per second) superimposed on them. A simplified diagram and an installation at one end of the terminals are shown in Figure 4-23 and Figure 4–24, respectively. The carrier signal normally operates the relays in such a manner as to keep the circuit breakers in a closed position.

When the transmission line is faulted, the carrier signal is interrupted, and the relays operate to open the circuit breakers. The system functions positively and effectively for several hundred miles. The carrier channel may also be used for other purposes; e.g., telemetering and supervisory controls that operate on coded impulses, for telephone communication, and for other unrelated business.

Microwave Relaying

In microwave relaying, the pilot systems are transmitted over microwave radio channels. This system is not subjected to line faults, and it may also accommodate several other separate functions. Because the transmissions are affected by line-of-sight limitations, several units may be required for long and tortuous lines. The intermediate units receive their signals and retransmit them to the next unit (referred to as *relay stations*).

Generally, the pilot schemes operate to keep the circuit breakers closed. When the line is faulted, the signals are interrupted, and the relays operate to open the circuit breakers. Similarly, if the pilot systems, including carrier and microwave systems, fail for any reason, the circuit breakers open to deenergize the lines and the transmission system "fail-safe."

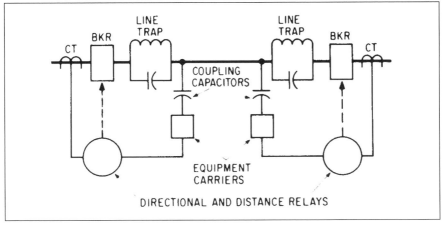

Fig. 4–23 Carrier pilot relaying equipment

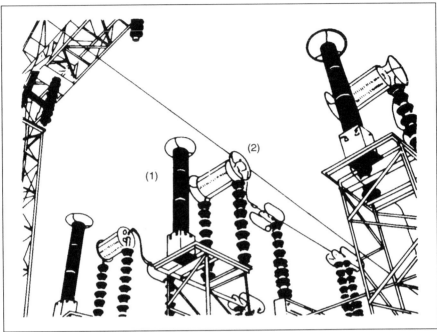

Fig. 4–24 Installation of the coupling capacitor (**1**) and the line trap (**2**) on a 345-kV transmission line

Ground Relay

Other schemes measure the flow of so-called ground current. In a transmission line, the currents flowing in each of the conductors are usually fairly well balanced in magnitude. The return or ground conductor carries little or no current. By measuring this directly, or by measuring each conductor and determining the difference, this ground current can be made to actuate relays when it exceeds certain predetermined values. Other more sophisticated schemes are sometimes used, but generally they employ one or more of the described basic ideas.

Figure 4-25 shows a relay using ground fault detection and illustrates the ability of this protection scheme to discriminate between load current and fault current.

Voltage Surges

Voltage surges can occur from lightning, switching, or fault conditions. Whatever the source, it is necessary to consider the effects not only on

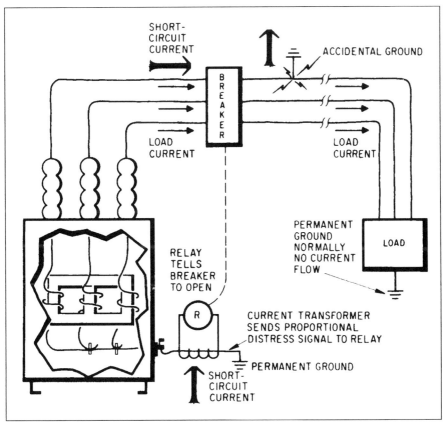

Fig. 4–25 Ground relaying

the transmission lines themselves, but also on the equipment that may be connected to them. These include switches, circuit breakers, transformers, generators, buses, regulators, and any other device that may be connected to them. Generally, these are situated at generating stations and substations. Much of this equipment is similar to and operates in the same manner as that found in distribution substations; however, several are of greater importance for the transmission system.

Figure 4-26 is a graphical representation of a traveling wave on a transmission line showing two of several possibilities of voltage surges occurring at points of discontinuity. (These could be an open switch, a transformer bank, or change of overhead to underground.) In (A), the characteristics of the point of discontinuity are such that the reflected wave is super-

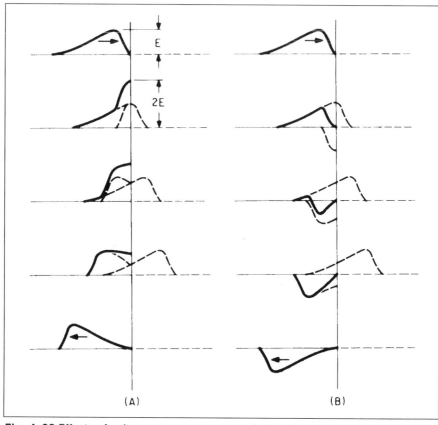

Fig. 4–26 Effects of voltage surges on a transmission line

imposed in the original surge voltage wave. The crest voltage is double the original value at the point of discontinuity. In (B), the characteristics of the point of discontinuity are such that the reflected wave is subtracted from the original wave.

Basic Insulation Level (BIL)

Since the operating voltages in a transmission system are relatively high, special attention is given to the insulation associated with the several parts of the system. This is true both in the lines as well as in the stations. The insulation here has to withstand not only the normally applied operating voltages, but the surge voltages explained later (see Fig. 4-27). Since

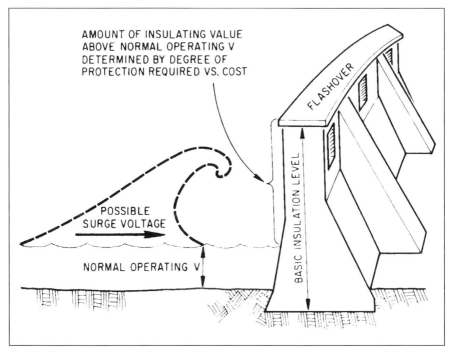

AMOUNT OF INSULATING VALUE
ABOVE NORMAL OPERATING V
DETERMINED BY DEGREE OF
PROTECTION REQUIRED VS. COST

FLASHOVER

BASIC INSULATION LEVEL

POSSIBLE
SURGE VOLTAGE

NORMAL OPERATING V

Fig. 4–27 The basic insulation level (BIL)

such insulation is expensive, it is desirable not to "overinsulate" any portion of the system unnecessarily. Hence, the insulation of the several components of the entire system must be coordinated, and a level of insulation to be provided decided upon.

Levels of insulation that will safely sustain the surge voltages, known as basic insulation levels (BIL), have been set up for electrical apparatuses, such as transformers, circuit breakers, and switches. These designated minimum levels may be from 2–3-1/2 times the normal operating voltage, depending on the degree of reliability desired. The minimum insulation level should prevail under wet conditions, and in general, should be the same for line insulation as for apparatus insulation.

At some point, usually at the bushing, the insulation value is made deliberately at the lowest value. This is arranged so that if a failure occurs, it will occur at a point that is readily accessible for repair or replacement.

Short-Circuit Duty

Again, because of the high voltages employed, when a fault or short circuit occurs on a transmission line, the amount of current that will flow will be inordinately great. This current flows not only through the transmission line conductors, but through all the apparatuses connected to it. This high-magnitude current produces magnetic fields of great intensity with corresponding great forces. These forces tend to pull conductors apart and damage lines and equipment. Hence, equipment, particularly circuit breakers, must be built rugged enough to withstand these disruptive forces.

The measure of the ruggedness of the equipment is referred to as the interrupting capacity or "short-circuit duty." It is generally expressed in kVA (thousand volt amperes) or mVA (million volt amperes). Hence, circuit breakers and other equipment are rated not only for their normal voltage and current carrying ability, but also their so-called short-circuit duty (e.g., 69 kV, 500 A, 500 mVA). As transmission grids become larger, with more generation interconnected, the fault currents and associated short-circuit duty become larger, making equipment more expensive.

Stability

When faults occur on a transmission grid supplied by two or more generators (see Fig. 4-28), the current flow to the fault will be proportioned to the distance (electrically) of the several generators. Thus, the generator closest to the fault will supply the greatest share (see Fig. 4-28a). As these heavy currents are imposed on the several generators, it will cause them to slow down, but not equally. Again, the one supplying the greatest share of current will slow down the most.

Consequently, these generators will no longer operate "in step." The one that slowed down least will attempt now to supply the other generators connected to the grid, which are now "bucking" it. This will cause it to slow down, and the process reverses. A rocking back and forth effect between generators will ensue. The generators then act very much as if they are connected together mechanically through a spring (see Fig. 4-28b).

If the fault is removed in time, this rocking effect will subside and the generators will snap back into step. If not, the effect will increase progressively until the current supplied by some generator exceeds its protective relay setting, and it will be disconnected from the grid. If the fault persists,

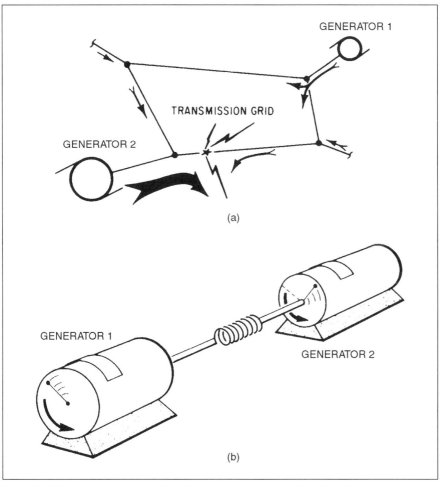

Fig. 4–28 Effects of faults on a transmission line connected to more than one generator. (**a**) Short circuit in a transmission grid. Generator 2 supplies a major portion of the short circuit current. This causes Generator 2 to slow down. (**b**) Generator 1 tries to assist. The effect is similar to a spring construction between generators, tending to keep them in step.

similar effects will cascade to other generators until all the units are disconnected, and a "blackout" results. This effect is generally referred to as a lack of stability or an unstable condition. Settings on protective relays and devices are designed to switch out the faulted section of the system as rapidly as possible, restoring the rest of the system to normal.

Cogeneration and Independent Power Producers

Large industrial or commercial consumers may sell their excess energy to utility companies. This can occur if the industrial or commercial consumer has an excess of steam energy that can be converted into electricity, or can generate electricity and has excess capacity. In addition to cogeneration installations, the independent power producer (IPP) has become an increasingly significant factor to the transmission system. The IPP constructs an electric generating station apart from the stations owned and operated by the utility system. Generating capacity from sources other than investor-owned and government systems has grown from 19,237 MW in 1970 to 55,188 MW in 1992.

This supply of power is accomplished by interconnecting the IPP installation to the utility transmission lines, usually through a substation installation. They must not only provide the necessary equipment and protective devices, but must coordinate them with the utility to which they may be connected. The power they put out may also be wheeled to other consumers or utilities. In any case, this portion of their operation must be placed under the control of the utility or pool system operator. Such installations may be considered as another of the transmission substations.

Review

- Substations receive energy transmitted at high voltage from the generating stations, reduce the voltage to a value appropriate for local use, and provide facilities for switching. They also provide points where safety devices may be installed to disconnect circuits or equipment in the event of trouble.

- Substations are generally located so that they will be as near as possible to the load center of the areas that they are intended to serve. Availability of land, cost, local zoning laws, future load growth, and taxes are some of the many factors that must be considered.

- Substations usually have two or more incoming supply transmission lines for reliability. Many of these stations are operated automatically, with control circuitry connected back to an operating center.

Substations may have an operator in attendance part or all of the day, or they may be entirely unattended. In some unattended substations, the equipment function is monitored remotely. If an alarm is signaled, a "roving operator" will be dispatched to operate that station.

• The voltage of the incoming supply is changed to that of the outgoing subtransmission of distribution feeders by means of a transformer. The nameplate on a transformer gives all the pertinent information required for the proper operation and maintenance of the unit.

• A busbar is a main bar, or conductor, carrying an electric current to which many connections are made.

• A regulator is a transformer with a variable ratio that maintains a transmission voltage at the specified level.

• Circuit breakers allow interrupting a circuit while current is flowing through it. Oil circuit breakers are the most common, but another type, called an air circuit breaker, puts out the arc by a blast of compressed air. Another type has its contacts enclosed in a vacuum or a gas, which tends to keep the arc from maintaining itself.

• Instrument transformers are used to measure large currents or voltages and to insulate a meter or relay from the circuit in which they are to operate.

• Relays provide protection for the components of a transmission system by providing tripping commands to the circuit breaker or breakers in an overloaded circuit.

• Differential relaying provides protection against faults on buses by comparing current supplied to the bus with current flowing from the bus. The currents should be approximately equal. A fault on a bus creates an imbalance that triggers a relay to clear both incoming and outgoing feeds from the bus.

- In carrier relaying, input and output currents are measured at both ends of a line. A pilot wire, microwave transmission, or other means are used to transmit these quantities to relays at both ends to clear the lines in trouble. Sometimes, the conductors themselves are used to transmit these signals.

- In ground relaying, ground current that is produced when the current flowing in the conductors is unequal can be made to actuate relays when it exceeds certain predetermined values.

- Basic insulation level (BIL) is the level of insulation that will safely sustain surge voltages. BIL must be considered when selecting devices such as transformers, circuit breakers, and switches. Minimum levels designated may be from 2–3-1/2 times the normal operating voltage, depending on the degree of reliability desired.

Study Questions

1. What is the function of a transmission substation? What other purposes may it serve?
2. What are some features of transmission substations?
3. What important factor influences the location of a transmission substation? What other factors are also considered?
4. What pieces of equipment may be found in a transmission substation?
5. What is the function of a substation transformer? Of a regulator?
6. What is the function of a circuit breaker? What types are there? How are they rated? Show a typical rating.
7. Why is it more important to provide protection for components of a transmission system than for a distribution system? How is this accomplished?
8. What are the functions of air-break switches and disconnects?
9. What is a relay? Describe three relay applications in transmission systems.
10. What is meant by the BIL of a piece of equipment?

5

Extra-High Voltage and Direct-Current Transmission

Increased Demand and Challenges

Increases in transmission voltages are due to many factors, but principally are the result of the need to provide ever-greater amounts of power over the same rights-of-way. This need for higher transmission voltages increases with increases in the cost of land (and its clearing) and decreased land availability. As power systems continue to be interconnected to provide greater economy and service reliability, that need becomes more imperative.

While such an increase in the voltage of a transmission line appears simple, it creates problems involving insulation, lightning protection, switching, and losses. Equipment such as circuit breakers, transformers, surge arresters, reactors and capacitors, towers and supports, and conductors (along with most other items of material) are affected. Corona losses, skin effect, and the effects of moisture, dirt, smog, and salt on insulators must be considered with other factors.

Audible noise problems, as well as problems with radio, television, and other communications, also tend to increase with overhead higher voltage transmission lines. Questions also are raised as to the effect of such high-voltage lines on human, animal, and plant life. These concerns apply as well to the effect on structures, vehicles, and other organic and nonorganic objects located or operating in their vicinity.

Undergrounding high-voltage transmission lines proves technically feasible but economically unfeasible, except in a few instances. (For example, this might be used for very short distances within such cities as New York.)

Direct-current (DC) transmission for long overhead or short underground or underwater distances is generally limited to intersystem ties of 250–1,500 MW capacities.

Operations

The capacity of transmission lines may be limited so that its outage or nonavailability will not inordinately affect the operation of the associated system or grid. This compares to the effect of having a large generator out of service. Further, for overhead lines that constitute the overwhelmingly majority of such lines, the maximum achievable transmission line loading is limited by the right-of-way available. It is evident, therefore, that environmental concerns (including beautification) and the desire for low-cost power are not easily reconciled.

Relationship of Line Capability and Voltage

For a given conductor, that is, a conductor with a fixed electrical resistance, the amount of electricity or current that may be carried by it economically is limited by one of two things. These are:

1. The permissible voltage drop or loss of electric pressure
2. The power loss that represents the heat dissipated from the line

Both must be considered in relation to either the transmission line voltage or total power transmitted. For example, permissible voltage drop may be 10% of the transmission line voltage, or permissible power loss may be 10% of the total power transmitted. Expressed in the form of an equation:

$$\text{transmitted power} = \text{current} \times \text{voltage}$$

or

$$P = I \times E_{line}$$

where

P   is the transmitted power
I   is the current
E_{line} is the line voltage

Then, if the voltage of a line is doubled, the current that will flow for the same power supply will be only half the original current.

Voltage drop and power loss may be expressed as follows:

$$E_{drop} = IR \text{ and } P_{loss} = I^2 R$$

where

I   is the current
R   is the resistance of the conductor
E_{drop} is the voltage drop or voltage loss
P_{loss} is the power loss in the transmission line

With half the current flow, the voltage drop or loss is also halved, and the power loss, with only half the current flow, will be only one quarter the original loss. Conversely, if the power loss can be maintained at its original proportional part, the transmitted power can be quadrupled. The voltage then can be doubled, and the amount of current sent through the same conductor can also be doubled. The power that this conductor now is capable of carrying, represented algebraically (as compared to the original), is:

$$P(\text{orig.}) = I(\text{orig.}) \times E(\text{orig.})$$

$$P(\text{new}) = 2I(\text{orig.}) \times 2E(\text{orig.}) = 4P(\text{orig.})$$

Hence, doubling the voltage of a line will increase its capability 4 times.

Extra-High Voltage

Many economic factors have contributed to the development of higher voltage transmission lines. Basically, the power that a line will carry increases as the square of its voltage increase. If the voltage is doubled, the power capacity is increased 4 times. More specifically, a 345-kV line will carry 9 times the power of a 115-kV line even though the voltage is only 3 times as high. A 765-kV line has 25 times the power capability of a 138-kV line.

When all the factors are considered, the 765-kV line is capable of carrying 4–6 times as much power as a 345-kV line over comparable distances. Put another way, the cost per megawatt of power transmitted is reduced considerably when a higher voltage is used. Tower lines and equipment are very large in size (see Fig. 5-1).

Cost Comparisons

In selecting the voltage of a transmission line, it is essential that the incremental differences in costs of all the other associated equipment be

Fig. 5–1 Guyed V-aluminum tower (170 ft high) with 765-kV line; each phase conductor consists of a bundle of four substrands (*courtesy Ohio Brass Company, a subsidiary of Henry Hubbell*)

taken into consideration. These include circuit breakers, transformers, switchgear, lightning protection, structures, buildings, rights-of-way, various control equipment, and measuring devices.

Cost consideration should include annual carrying charges (annual losses of energy from all sources). These are influenced by the different voltages under consideration and must be taken into account. Results should, however, be tempered with practical considerations, such as the availability of skilled workers and potential access problems. It is very important to consider environmental protection, the appearance of the line, and the effect it might have on the community.

Comparisons can be made of costs with alternating-current (AC) systems of similar voltages and load-carrying abilities. While costs of terminals and other associated equipment are essentially the same, line costs can be substantially lower (see Table 5-1).

Mine-Mouth Generation

The use of extra-high voltages makes practical mine-mouth generation (see Fig. 5-2). Here, instead of transporting coal to the generating plant, it is cheaper to generate the power right at the mine and then transport the power by transmission line. Similarly, other remote sources of power, as hydroelectric plants, are made practically available to major consumer centers. Because of insulating material limitations, voltages produced by generators are restricted to about 20 kV. The AC output, however, is readily

Table 5-1 Economic Comparisons for Extra-High Voltage Transmission (in percent)

Equipment for AC Transmissions	450 kV	765 kV	1,000 kV
Circuit breakers/unit	100	200	400
Autotransformers/kVA	100	125	150
Reactors/kVAR	100	120	140
Series Capacitors/kVAR	100	110	120
Line/mile	100	150	210
Equipment for DC Transmission		±375 kV	±900 kV
Terminals, including rectifiers/kW		200	220
Line/mile*		50	200

*Compared to AC figures.

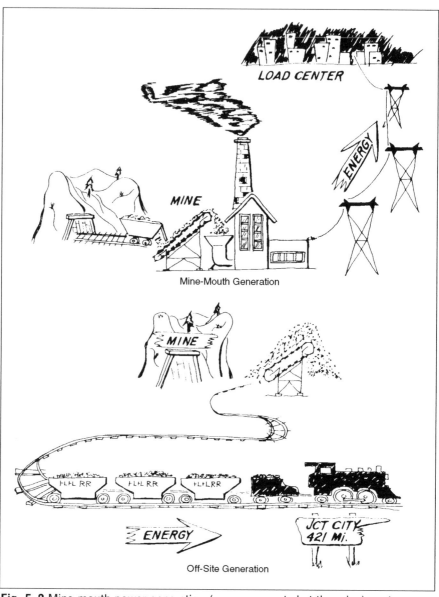

Fig. 5–2 Mine-mouth power generation (power generated at the mine) vs. transportation of coal to generate electricity elsewhere

changed by means of transformers, so the generated voltages are easily stepped up to transmission voltages.

Equipment

Circuit Breakers

Circuit breakers are becoming more complex. Because the higher voltages tend to have longer arcs that persist for longer periods of time during the disconnecting phase of their operation, greater attention is paid to the insulating arc-quenching medium. Use of sulphur hexaflouride (SF_6) gas under pressure has solved the problem of arcing, at the same time providing insulation qualities that are roughly 100 times better than oil. The pressure applied to the gas, however, aggravates the possibility of leakage.

The sudden interruption of current at these high voltages creates voltage surges greater in magnitude than line voltages. These surges not only are applied to the breaker conductors, but travel out along the lines until they dissipate or are bled to ground through surge arresters. The effect of this phenomenon is to require greater insulation to be specified at critical points in the breaker, the lines, and in other equipment that may be connected to them. The interrupting duty of such breakers becomes greater as values of voltage and current become larger, requiring much greater mechanical reinforcement of the component parts. The size of the breaker may be inordinately large.

Transformers, Reactors, Capacitors

Similar problems of higher voltages, surges, and mechanical stresses imposed on component parts apply to these pieces of equipment as to circuit breakers (detailed in previous section). Structural reinforcement of coils, cores, and supports are necessary to withstand the extraordinarily high currents that may flow when faults or short circuits occur anywhere on the line. As no arcs are to be quenched, oil or askarels are satisfactory for insulation purposes. Because of possible voltage surges causing flashover, bushings may be proportionately larger than those for lower voltage equipment.

Lightning or Surge Protection

Voltage surges caused by lightning or switching are more apt to result in flashover across insulator strings and to the tower or support structure. They involve problems of clearances and insulator swing geometrics.

Lightning striking at or near an overhead transmission line creates surges of voltage that travel along the line. This action is the same for all overhead lines exposed to lightning. Extra-high-voltage transmission lines differ in that the line current that flows in the flashover tends to be much greater because of the greater line voltage. Obviously, this greater current imposes greater stresses on the equipment through which it may flow, with possible damage to the equipment. This is especially true of the circuit breakers called upon to interrupt such currents.

Not only are a greater number of lightning or surge arresters installed along the line, but greater effort is made to reduce ground resistances to enable the surge voltage to be dissipated as quickly as possible.

While the voltage surges caused by switching may not be as great as those caused by lightning, the same general conditions result as those described, and the solutions are the same. To aid in reducing damaging flashovers, extra insulators may be added to the strings than would normally be required. The longer string of insulators allows the conductor attached to it to have a greater radius in its sway. This calls for greater clearances of the conductor from the tower or supporting structure. Furthermore, the string must be carefully placed so that the swinging conductor will not make contact with adjacent conductors.

For the principle of operation of surge arresters, the several types, and their application, refer to chapter 2, "Overhead Construction."

Underground Transmission

The capacitance or condenser effect of high-voltage cable absorbs energy and limits the amount of useful energy as well as the distance over which it may be transmitted. To reduce this capacitance effect, compressed gas (sulphur hexaflouride) at 50 pounds per square inch pressure is employed as insulation in extra-high-voltage cables. While satisfactory for this purpose, it introduces problems of maintenance (leakage and repairs). Nevertheless, this form of insulation can withstand the great stresses of high voltages under varying temperature and current loading conditions.

Superconductors

Liquified gas—such as nitrogen circulating within the conductors to maintain extremely low (cryogenic) temperatures—makes possible conductors having very low resistance. For example, if aluminum is cooled to the temperature of liquid nitrogen (-320°F), its electrical resistance is approximately 1/10 the value at normal temperatures. (Its mechanical properties change at such low temperatures.) However, necessary refrigeration and associated equipment to maintain the liquidity of the gas for the continuous removal of heat requires special manufacture and use of certain materials, resulting in an extremely costly installation.

Certain comparatively rare metals, such as Niobium, if cooled to temperatures approaching that of liquid helium (-425°F), lose all resistance to the flow of direct current. Such cables are termed superconductors, as compared to the cryogenic cable mentioned. The difference is in the temperatures and materials employed. Economically, such cables could become practical, offering the possibility of substantial reduction in cost in the high-power capability range (5,000 mVA and greater).

DC Transmission

Comparison of AC and DC

In a circuit, the maximum voltage permissible is fixed by its insulation. In DC systems, the maximum voltage is applied throughout the entire time the flow of electricity takes place. In AC systems, the maximum voltage is applied only a portion of the time. If the values are averaged (from 0 to maximum and back to 0), it would be found that this average value is 70.7% of the peak value (see Fig. 5-3). Hence, the capacity of a two-wire AC system is limited to only 70.7% that of a two-wire DC system having the same insulation and the same conductor size. This must be considered in order that the insulating value of the insulation not be exceeded.

DC Transmission Features

Many benefits can be realized by using DC for high-voltage transmission. In AC circuits, the effective voltage is 70.7% of the peak value the line carries; in DC circuits, the effective and peak values are one and the same. Hence, for a particular voltage rating, the DC circuit requires only 70.7% of the insulation required by the AC circuit. Conversely, with the same size cables and the same insulators, a DC line can carry about 40% more power. Further, the conductors are not subject to skin effect, although corona discharge continues to be a problem.

Since the circuits are not subject to alternating magnetic fields, no inductance within a conductor or between conductors is generated (except at the moments of energization and deenergization). Hence, voltage and

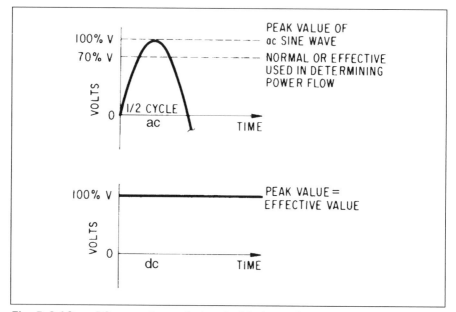

Fig. 5–3 AC vs. DC power transmission. In AC, the peak voltage must be used in calculating the insulation required from conductor to grounded supporting structure. This value is higher than the effective value of the DC system shown at the bottom. The DC system utilizes the maximum voltage to ground to transmit power.

energy losses are reduced. Further, as will be described later, fewer conductors are employed in a DC circuit as compared to an AC circuit. Generally two or three are used in DC as compared to four in AC. In some cases, where "ground" is used as a return circuit (such as a body of water), only one DC conductor may be used.

Advantages

High-voltage DC permits the use of higher effective voltages for a given number of insulators in a string and a given length of flashover distance between the conductors and supporting structures. Leakage and corona losses are thereby decreased.

Since there is no alternating magnetic field around the conductor (except at the time of energizing and deenergizing), there are no inductive or capacitive effects. In AC systems, these often require corrective measures. The power factor is unity, a fact that alone accounts for considerable reduction of transmission losses. This is especially important in underground cable systems. Similarly, the lack of a moving magnetic field reduces the stresses on circuit breakers, reducing their interrupting duty requirements.

Another important advantage of DC transmission is the ability to interconnect separate transmission systems (of the same voltage) without the necessity of first synchronizing them (see Fig. 5-4). This essentially eliminates the stability problems associated with interconnected AC transmission systems.

DC lines may consist of few conductors (use of only one or two conductors compared to three or more for AC systems). Consequently, tower and supporting structure requirements as well as those for rights-of-way are relatively less and are reflected in lower line construction and maintenance costs.

In addition, DC cable circuits have some special advantages over AC. For long cable circuits, DC has a lower investment cost because fewer individual cables are needed. DC does not have the problems of high amounts of reactive power that are produced in an AC cable. Also, DC cables have lower losses than equivalent AC cables (in long AC cables, shunt reactors are usually required to counter the capacitance effect of the cable itself).

Disadvantages

It is generally uneconomical to tap DC lines, and tapping is generally avoided. Line tapping requires conversion to AC before lower (or higher) voltages can be transformed to desired values. If the desired voltage is DC, conversion from AC back to DC is also required (see Fig. 5-5).

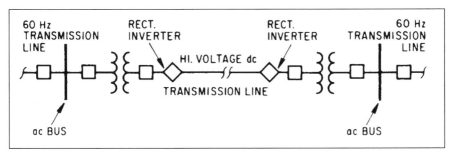

Fig. 5–4 DC transmission used over very long distances. The two AC buses may be hundreds of miles apart and do not have to be synchronized, or in phase, to permit power to flow between systems.

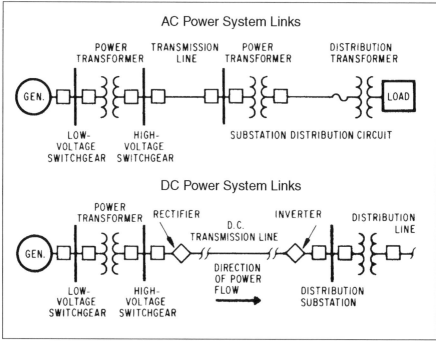

Fig. 5–5 Arrangement of AC–and DC–power system links

Changing Voltage—DC Systems

To raise or lower voltages in a DC system, it is necessary to go through the same procedure described for tapping of such circuits. Costly terminal equipment is necessary to convert power to AC at transmission and distribution substations to incorporate DC lines. Further, there are challenges associated with tapping DC lines or changing values of DC voltages. It has not yet proven feasible to control power flow and protect the circuit in the same fashion as a tap-off or change in voltage of an AC circuit.

Rectifiers

Conversion of AC power to DC power (and vice versa) is accomplished through rectifiers. For the amounts of power to be transmitted in such extra-high-voltage lines, mercury arc rectifiers, sometimes referred to as mercury valves, are used. The number employed depends on the amount of power to be converted.

Mercury arc rectifier installations are usually large and require periodic maintenance. Moreover, the stability of the arc is subject to momentary dips in the incoming source that may cause the arc to be extinguished, necessitating restart. Later installations employ solid-state rectifiers, sometimes referred to as thyristor valves, replacing the mercury arc converters. While these, too, are costly, they require little or no maintenance.

In short, high-voltage DC is suitable almost exclusively for transmission of large blocks of power over long distances in point-to-point transmission. The cost for DC lines is less than for AC lines. However, there are costs associated with the converter stations necessary for the conversion of AC to DC for transmission and then back to AC transmission for short lines. Economically, a break-even point in costs for high-voltage DC compared to extra-high-voltage AC appears to be at line lengths of about 400–500 miles. At this point DC appears economically feasible. As for underground cable circuits, this break-even point appears to be about 40–60 miles.

History

The earliest transmission lines were short, low-voltage, low-power, DC runs, confined to European countries. They consisted of a series circuit, known as the Thury system (see Fig. 5-6). Here a number of DC generators were connected in series to develop the high voltage of the transmission line, carrying a constant value of current.

At the receiving end were a number of motors connected in series, located at one or more substations, with the current flowing in each one having the same value. These motors, in turn, drove DC generators whose output served low-voltage DC distribution requirements. Variations in load, which necessitated changes in the voltage to vary with the load, were taken care of by short circuiting one or more generators as well as motors at the substations.

The vast increase in demands for electricity, especially with the advent of World War I, were met by the rapidly developing and expanding conventional AC systems, employing the transformer (with no moving parts). Thury systems were finally supplanted by conventional AC systems, although the last ones persisted until the early 1930s.

DC transmission is emerging as a transmission tool for large-capacity links over long distances, particularly for interconnections of dissimilar

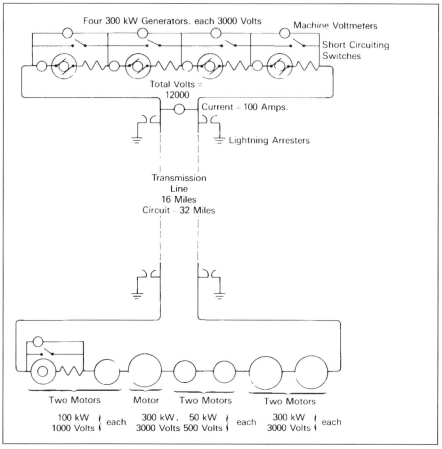

Fig. 5–6 Diagram of connections—Thury series system

areas. India, for example, has a 1,500-MW, 500-mile link. There also is a 2,000-MW link between eastern Canada and northeastern United States. Thyrister converter stations and new insulation material for cable circuits are making DC transmission links possible both in the United States and overseas. It is expected that by 2005, the total capacity of DC links will exceed 14,000 MW.

Review

- Extra-high-voltage (EHV) transmission often permits transmitting large blocks of power more economically than lower voltage transmission.

- Effective voltage of an AC system is only 70.7% of the maximum voltage of the AC sine wave.

- EHV transmission is ideal for "mine-mouth" power generation; the power-generating station is located right at a coal mine. This eliminates having to transport the coal to a generating station miles away.

- Load-carrying ability of transmission lines varies as the square of the operating voltages; doubling the line voltage results in quadrupling the line capacity.

- EHV transmission lines have serious effects on the design and construction of related equipment, such as circuit breakers, transformers, reactors, capacitors, lightning or surge arresters, conductors, towers and supporting structures, etc.

- EHV lines are subject to insulator flashovers and can subject equipment to large mechanical stresses.

- Flashovers caused by lightning or switching may result in greater flow of follow current that may cause greater fault current to flow, imposing greater stresses in equipment.

- Liquified gases at extremely low temperatures (in the −400° F range) circulate within conductors, causing their resistances to approach 0. These are often referred to as cryogenic cables and superconductors.

- The major advantages of DC transmission over AC transmission are:

  1. Higher power capability for the same size conductor and insulation
  2. Elimination of skin effect and inductances (loads are delivered atessentially unity power factor)
  3. Fewer conductors required for transmission

- The major disadvantage of DC transmission is the difficulty in stepping voltages up and down, requiring expensive equipment, including rectifiers.

Study Questions

1. What is meant by extra-high-voltage transmission?
2. What is the relationship between line voltage and capability?
3. What are three main considerations in selecting the type and voltage of a transmission line?
4. What are the advantages of EHV transmission?
5. What are the disadvantages of EHV transmission?
6. What is meant by "mine-mouth" generation? Where is it used?
7. What are the advantages of DC transmission compared to AC transmission?
8. What are some of its disadvantages?
9. What are superconductors? What are their advantages? What are some of their disadvantages?
10. How might AC and DC systems be interconnected?

6

Dynamic Capabilities

Conventional procedures used to calculate current-carrying capabilities of transmission-class overhead and underground lines depend on complex formulas and on conservative assumptions of conditions under which lines are operated. In transmission lines, normal ratings may not be as limiting a factor as the emergency ratings. Transmission systems are usually designed to carry the load even when one or more lines are out of service under a contingency condition, or during the loss of a generating unit or interconnection.

Transmission-line capability and reliability have also become increasingly important because few new transmission lines have been built in the United States since the 1970s. As shown by Table 6-1, bulk transmission line additions (220 kV and above) declined by more than 60% from 1975 to 1990. Even with this trend, economic and environmental restraints in the electric utility industry dictate that planners and operators make better use of existing system facilities.

Table 6–1 Average bulk transmission lines added (in the United States)

5-Year Period Ending	Avg. Net Miles Increase/Year (for 5-Year Period)
1975	6,467
1980	4,767
1985	4,051
1990	2,269

Source: EEI historical statistics

An Emerging Technology

Conservative assumptions in calculating circuit ratings include conductor temperature, ambient temperature, wind speed, subsurface conditions (soil characteristics and moisture content), historical load data, and insulation and conductor condition in service. Many operators and planners use these calculations for certain periods of the year (often summer and winter). The result is lines being underutilized or overloaded when actual conditions fall short of or exceed assumed conditions.

The challenge, then, is to measure actual operating conditions, automate the data-gathering, and integrate the information with accepted rating calculations. This can be used to produce "real-time" capability ratings (compared to "book" ratings) on a timely basis. In this way, technicians can operate the system safely and with full utilization of its capability (see Fig. 6-1).

Dynamic Rating Systems

History of Dynamic Rating Systems

In the early 1970s, L. Fink recognized the potential benefits of real-time knowledge of the actual data of a transmission circuit's thermal state. He recommended the creation of real-time monitoring and computer programs for establishing circuit capability. In the late 1970s, the Cable Monitoring and Rating System (CMARS) was developed through Department of Energy sponsorship. M. Davis made a similar approach for overhead lines.

In 1975, Underground Systems Incorporated (USi) was formed to supply real-time capability system components and technical evaluation for a CMARS demonstration project on cable circuits owned by Public Service and Gas in New Jersey. It was the first real-time transmission monitoring and rating system ever installed and produced valuable information before it was taken out of service when the demonstration period ended.

In the mid-1980s, a commercial, real-time monitoring and dynamic rating system for transmission cables, UPRATE™, was developed by USi. It was first installed on the Boston Edison system followed by installations at Long Island Lighting Company, PSE & G, Consolidated Edison of New York, and the New York Power Authority (see Figs. 6-2 and 6-3). The real-time systems were augmented by systems for overhead transmission lines and substation transformers in the late 1980s.

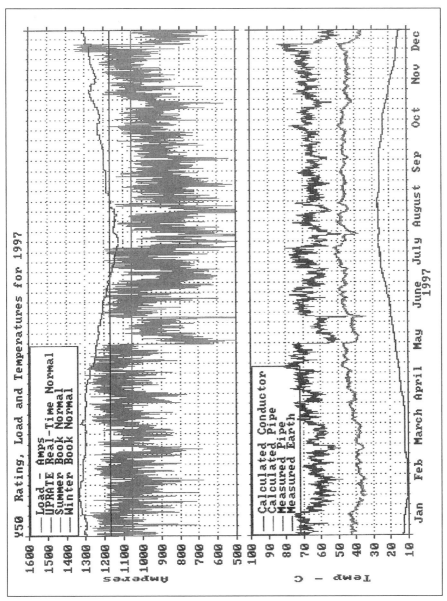

Fig. 6–1 Book vs. real-time capability ratings (*courtesy USi*)

Y50 345 kV Intertie - UPRATE® System

LILCO Operations Center

Fig. 6–2 System operator console (arrow points to real-time monitor) (*courtesy USi*)

Dynamic Rating System Design Requirements

Real-time dynamic ratings must be accurate for all load, environmental, and equipment operation conditions. This is essential to ensure that cable and equipment are not overloaded and helps to build confidence in those using real-time ratings for normal or emergency conditions. It also helps those who plan system reinforcements or additions.

Three elements are required to ensure the accuracy of real-time ratings: modeling, measurement, and parameter identification. These elements are common for all of the equipment in the power delivery system including overhead transmission lines, underground transmission lines, and substation equipment.

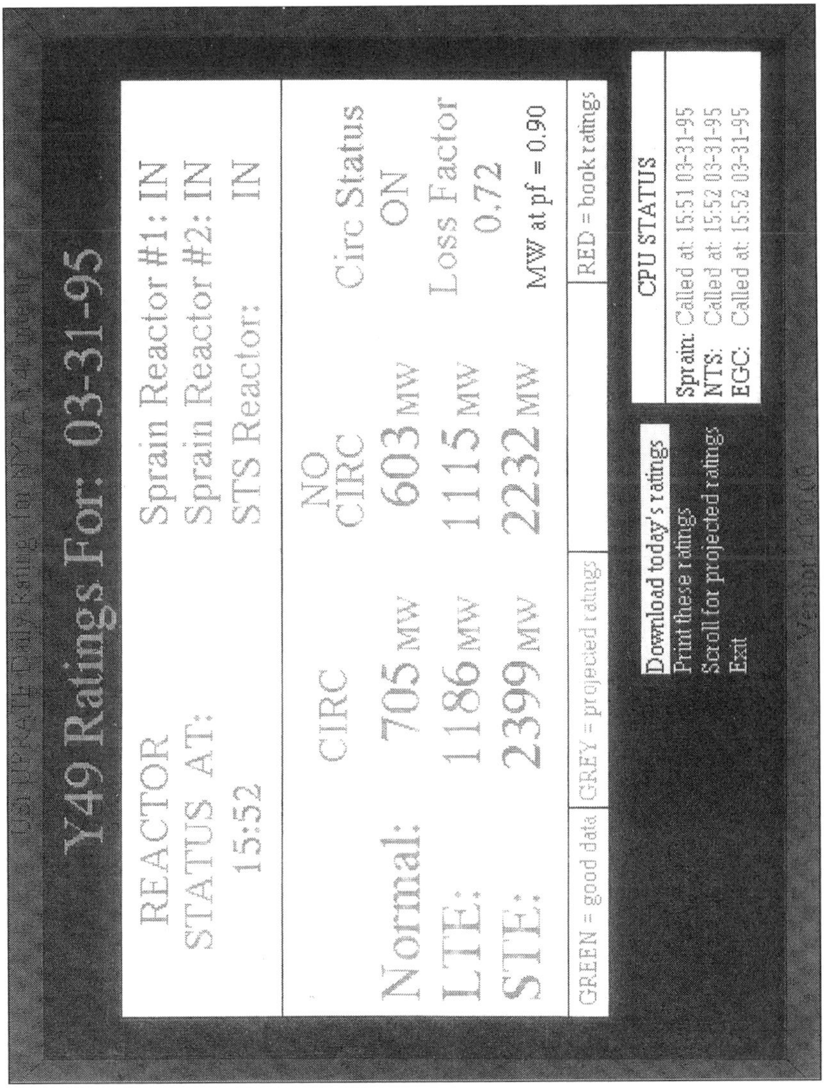

UPDATE Daily Rating for NY2: Negligible

Y49 Ratings For: 03-31-95

REACTOR STATUS AT: 15:52			Sprain Reactor #1: IN
			Sprain Reactor #2: IN
			STS Reactor: IN
	CIRC	NO CIRC	Circ Status ON
Normal:	705 MW	603 MW	Loss Factor 0.72
LTE:	1186 MW	1115 MW	
STE:	2399 MW	2232 MW	MW at pf = 0.90

GREEN = good data GREY = projected ratings RED = book ratings

Download today's ratings
Print these ratings
Scroll for projected ratings
Exit

CPU STATUS

Sprain:	Called at: 15:51 03-31-95
NTS:	Called at: 15:52 03-31-95
EGC:	Called at: 15:52 03-31-95

Fig. 6-3 Screen display of real-time ratings (*courtesy USi*)

An accurate dynamic thermal model of the equipment that operates in real-time is the basis of an accurate rating system. The thermal model must correctly describe the equipment's thermal performance, both during normal function and during emergency situations. For example, thermal models for transmission lines track problems arising from heat transfer in pipe-type cables or from sag clearances on overhead lines. They provide temperature and rating information in the event of an operational problem. The model must allow quantification of the actual equipment capacity and, in addition, operational condition and environmental parameters critical to the stability of the equipment.

Second, equipment must be installed to measure key parameters on a real-time, continuous basis. Parameters that drive the model, such as load, temperature, wind speed, etc., and the operational status of equipment must be measured. They are needed to identify the thermal characteristics of the equipment that relate to its thermal response (see Figs. 6-4, 6-5, and 6-6).

Third, parameter identification algorithms must be used continually to evaluate and update key, uncertain thermal parameters. For all power delivery components there are key parameters that are "uncertain" in that they can and do change. However, they can have a significant effect on the temperature and ratings of the equipment. For example, in underground transmission cables, these parameters include earth resistivity and earth ambient temperature. For overhead lines the uncertain variable is wind.

Benefits/Typical Performance

All of these dynamic rating systems provide a variety of benefits. These include rating increases, improved safety in the use of ratings, identification of environmental problems, and equipment diagnostic information that can reduce the incidence of future failures. Dynamic ratings systems can yield enormous operational savings when real-time ratings score higher than book ratings. This allows use of economic generation within the system.

Open-access transmission and marketing of power make more effective use of interconnections to limit serious system problems involving possibility of blackouts (loss of service to large areas). In some emergency situations, if the system operator follows book ratings, he may have to interrupt load or lose the entire network. With real-time capability, he finds that capacity limits are usually higher than the book ratings, particularly for emergency ratings. Consequently, he may not be forced into a situation in

Fig. 6–4 Installation of sensor donut on OH 230-kV line (*courtesy USi*)

Attachment of Thermocouples on Trailing End

Fig. 6–5 Installation of thermocouples on a new 138-kV cable (*courtesy USi*)

which he must either drop load or lose an entire area. System planners may find they do not need to reinforce a transmission line or construct a new one to avoid exceeding the book rating of a transmission component.

Actual commercial use of dynamic rating systems, such as USi's Uprate™, has resulted in useful data concerning underground cable and overhead transmission lines.

Underground Cable

Dynamic ratings on cable systems typically boost short-time emergency cable ratings 40–110% and normal ratings 15–25% compared to book ratings. Transmission capacity is not only increased but made safer, because the system ratings are based on actual cable component temperatures and earth thermal conditions.

Fig. 6–6 Real-time data CPU in substation (*courtesy USi*)

For example, cases have occurred in which the utility book rating was based on an assumed earth *rho* (measurement of soil thermal resistance) of 60, when in fact the earth rho at the monitored cross section was more than 100. Alerting engineering/operations to take corrective action may well have averted an expensive cable failure from the heat effect on the insulation.

Overhead Transmission Lines

Actual utility data show dynamic ratings usually exceed book ratings 70–90% of the time for well-defined periods of the day. Further, due to typical wind conditions during midday, they often exceed the book ratings during the hottest time of the day when load requirements are greatest. During periods when winds are lighter than expected and dynamic ratings are lower than book ratings, the dynamic ratings are more accurate to maintain safe conductor temperatures for sag and clearance requirements.

Integrated Power Delivery Data and Dynamic Rating Systems

Real-time monitoring and dynamic rating systems for power delivery components can be integrated to provide real-time circuit ratings for all of the load-carrying equipment on a circuit or segment of the system. To determine this circuit rating, the ratings of each component must be established and the lowest rating selected in real time. This becomes the circuit rating. This lowest-rated component can then either be replaced, reinforced, or investigated for means of increasing the rating.

System-wide implementation of real-time monitoring and dynamic rating for power delivery allows users to easily access the data for a variety of uses. (Data are made available through a power delivery applications server.) These uses include rating and temperature information for system planning and engineering and diagnostic status for maintenance and substations operations. It also provides real-time dynamic ratings for system operations and power pool/ISO organizations in the format required.

From an economic standpoint, utility studies have shown that deferred installations and economic dispatch improvements can pay back investments on dynamic systems in less than a year. They can provide additional generation revenue for future power delivery marketing and power generation conditions prevalent in today's competitive energy business.

Appendix **A**

Environmental Considerations

Many environmental considerations are taken into account in the planning, design, construction, maintenance, and operation of transmission lines and associated substations. A few examples follow to illustrate the scope and magnitude of such considerations.

Safety

Overall, perhaps the prime considerations are the measures taken to protect people from injury. In addition to maintaining certain minimum clearances of hazardous objects from the general public, there are other important measures. These include:

- Barriers, fences, and walls around potentially dangerous areas
- Locks on doors, gates, operating handles, and other accesses to facilities and equipment
- Interlocks on critical equipment and devices to prevent incorrect operation and accidental energization and deenergization of facilities that could endanger workers and the public
- Grounding of fences, towers, and other metallic structures accessible to the general public that could be accidentally energized

- Smoke and fire alarms to warn the public of impending possible danger
- Measuring devices installed throughout a region to monitor pollution of the surrounding atmosphere

Aesthetics

Transmission lines are routed to avoid areas of particular interest. Such areas might include:

- National monuments or regional parks
- Areas of population density or recreational importance
- Areas where people congregate for special events
- Playgrounds
- Buildings and other structures of historical interest
- Areas for the preservation of the integrity of national resources and ecological preserves, including animal sanctuaries and forest reserves

Much attention has been directed toward making transmission lines and associated substations less obtrusive and more pleasing (or less displeasing) to observers. Poles and support structures have been gracefully designed to blend with their surroundings (see Figs. A–1, A–2, A–3, and A–4). Similar treatment is applied to insulators and equipment mounted on the structures. For lower voltage lines (up to 69 kV), lines are constructed without cross-arms to improve appearances (see chap. 1).

Rights-of-way are cleared of unsightly growth, and landscaping is provided in some areas; in other areas, flowers and other plants are used to create pleasant garden effects. Substations are landscaped to conceal or beautify the enclosures of equipment (see Figs. A–5 and A–6). In some instances, the station is contained within structures made to conform and blend in with the neighborhood (see Fig. A–7).

When other means have proven impractical, lines are placed underground for short distances through the areas affected, including cable adits and exits from substations.

Pollution

Sight is not the only sense to which attention is paid. Steps are taken to eliminate or abate disagreeable sounds. Acoustic barriers are installed,

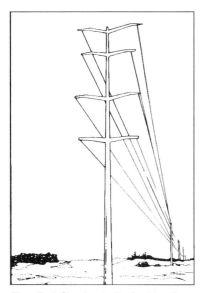

Fig. A–1 Single-shaft, double-circuit, 345-kV line (*courtesy FL Industries, MN*)

Fig. A–2 H-frame, double-circuit, 500-kV line (*courtesy FL Industries, MN*)

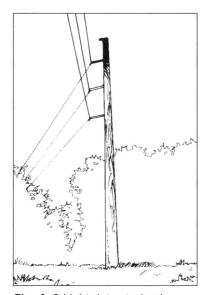

Fig. A–3 Light-duty steel pole (*courtesy FL Industries, MN*)

Fig. A–4 Changes in 138-kV line appearance: 1987, left, and 1937, right (*courtesy Long Island Lighting Company*)

generally at substations, to keep annoying sounds confined within limited areas. In some instances, devices producing sound frequencies to cancel those produced by energized equipment are installed to lessen this form of nuisance.

There is a possibility of pollution of surface and subsurface water sources from oil spills from electrical equipment (transformers, circuit breakers, etc.). Dams and barriers are carefully designed and constructed around such equipment to prevent this from occurring.

Ecology

The preservation of wildlife refuges and sanctuaries, including endangered species, by the rerouting of facilities has been discussed previously.

Fig. A–5 Green Acres substation. Heavy landscaping matches that typical of the area. Growth is monitored to maintain appearance, shielding structures from view as much as possible. (*courtesy Long Island Lighting Company*)

Fig. A–6 Atlantic Beach substation. Low-profile substation has fencing and shrubs typical of the neighborhood. Structures are painted in colors defining their type and function, with special emphasis on hazardous areas. Major equipment is colored to blend in with the environment to be as inconspicuous as possible. (*courtesy Long Island Lighting Company*)

Fig. A–7 Seaport substation in Manhattan. Walls around equipment are made to look like other industrial buildings in the area. (*courtesy Consolidated Edison Company*)

Other provisions are made to protect some wildlife from self-destruction (as well as to prevent interruption of service). These include shields placed on the lower parts of poles, towers, and other structures to keep such animals from contact with energized conductors. Similar guards are installed on equipment bushings to serve the same purpose.

In some instances, usually where ospreys, eagles, and other birds with large wingspans exist, nests are mounted on separate structures a short distance away from the transmission lines. These are meant to keep such birds from nesting on or in the line structures, where they may come in contact with energized conductors.

Hazardous Materials

Hazardous materials usually are avoided; where they may exist still, steps are taken to eliminate them. Where insulating oils and askarels contain polychlorinated biphenyl (PCB), steps are taken to replace the fluids (where necessary the entire equipment itself), or clarify them. (PCB is considered a teratogen.) Again: Disposal falls under "regulated hazardous waste," per the federal Toxic Substance Control Act (TSCA).

Similarly, where there is asbestos insulation or noise barriers that contain asbestos, steps are taken to remove this material or replace it with other nonhazardous materials (see chap. 4).

Electromagnetic Radiation

In the case of higher voltage transmission lines, research continues on the effect of their strong magnetic fields on human, animal, and plant life in their vicinity. As is known, a voltage is induced in a conductor cut by a magnetic field, its magnitude in part depending on the strength of the magnetic field. People, animals, and plants constitute conductors of sorts. They have voltages induced in them that cause currents to flow within them as "eddy" currents, as well as between them and the ground or other objects with which they may come in contact.

The effect these currents have on the biological system, and especially on the nervous system, appears to be minimal or nonexistent, although the length of time of such exposure may play an important part. Meanwhile, tentative standards are being drawn calling for minimum widths of rights-of-way in accordance with the voltage of the transmission lines. For example, some states have legislated a minimum right-of-way width of 350 ft—or 171 meters (m). Additional width is provided to maintain a maximum magnetic field strength of 1.6 kV/m from conductor to the area or structure in its vicinity, applied to the shortest distance between them.

These observations apply to both AC and DC lines. Magnetic fields produced by the former alternate in direction according to the frequency of the circuit. This constant to-and-fro motion produces alternating current on whatever bodies that are cut by the magnetic fields. In DC transmission, the magnetic fields produced are stationary (except when first energized and again when deenergized). However, objects in motion in their vicinity cut the magnetic field, producing a direct current within them.

Tentative "wire codes" have been established by some utilities that correlate the number of wires and their diameters and voltages to specific "safe" distances from vulnerable objects. Wire codes are modified by other factors. These include pollution effects on the lines, population density in the vicinity of the line, traffic in the area, the shielding effect of other structures, and other factors that may influence possible current flow.

* * *

These are only a few of the measures taken to improve and maintain the quality of the environment that may be influenced by transmission lines. They are an indication, however, of the scale and scope of the considerations taken into account that affect transmission lines from the planning stages to their final construction and operation.

Guide for Uprating Transmission Structures For Higher Operating Voltages

Transmission lines can have their capacities uprated to accommodate increased loading to utilize existing rights-of-way. This may be accomplished by installing larger conductors on existing structures, increasing operating voltages, or both. In general, uprating existing lines creates fewer environmental, economic, and public relations problems than obtaining rights-of-way for new facilities. In such uprating, however, reconstruction and replacement of structures as well as revisions and beautification of existing rights-of-way may be necessary.

The guide details the handling of all of the elements in the transmission line to be uprated. It lends itself admirably to working out the details for obtaining and preparing rights-of-way as well as those associated with the construction of a new transmission line.

Standards and methods employed are those of the Rural Electrification Administration (REA) of the U.S. Department of Agriculture, to whom thanks are due for extending this courtesy. While the standards and methods of REA may differ in some details from those of other public and private utilities, they do represent a consensus of the utility industry. They should prove useful for the purposes indicated. This is especially appropriate when it is remembered that important REA activities include the planning, design, construction, maintenance, and operation of high-voltage, large-capacity, very long cross-country transmission lines.

Additional information, specifications, and other material may be obtained from REA offices.

Case Study: 69-kV to 115-kV Line Uprating

Introduction

This analysis provides a brief summary of a 69-kV to 115-kV transmission-line uprating study. The original 69-kV line was constructed on the standard REA TS-1 single-pole structure (see Fig. B-1). The uprated 115-kV line structure was constructed on the TH-1AM structure (see Fig. B-2).

Analysis

The original 69-kV line was designed and constructed in 1951–1952 under the safety guidelines of the fourth edition of the National Electrical Safety Code (NESC) and the then-existing REA "Transmission Line Design Guides." The uprated 115-kV line was designed in 1979 in accordance with the 1977 edition of the NESC, the 1972 REA Bulletin 62-1, and July 1978 "File With" REA Bulletin 62-1 vertical clearance requirements.

A tabulation of the original and uprated design criteria is provided in Table B–1.

Table B–1 Original vs. Uprated Design Criteria

Design Item Specified	Original Design Summary	Uprated Design Summary
Conductor	#1/0 ACSR 6/1	477 MCM 26/7 ACSR
Shield wire	3/8" H.S. Steel	3/8" H.S. Steel
Loading	3/8" Ice & 4# Wind @ 0°F	3/8" Ice & 4# Wind @ 0°F
Ruling span	500 ft	380 ft
Voltage	69 kV	115 kV
Basic poles	50 ft Cl.3	(50 ft Cl.3 & 55 ft Cl.3) 1 each

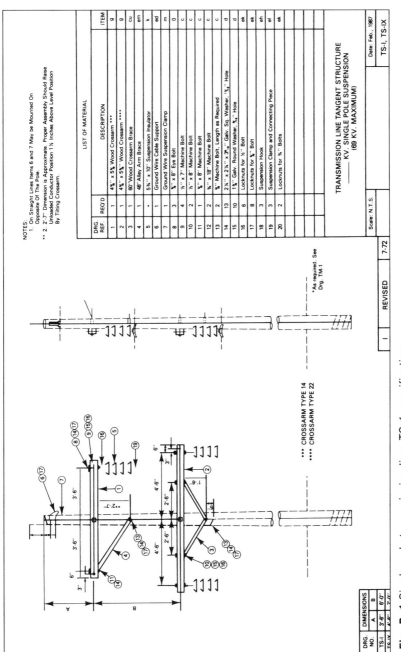

Fig. B–1 Single-pole transmission line, TS-1 specifications

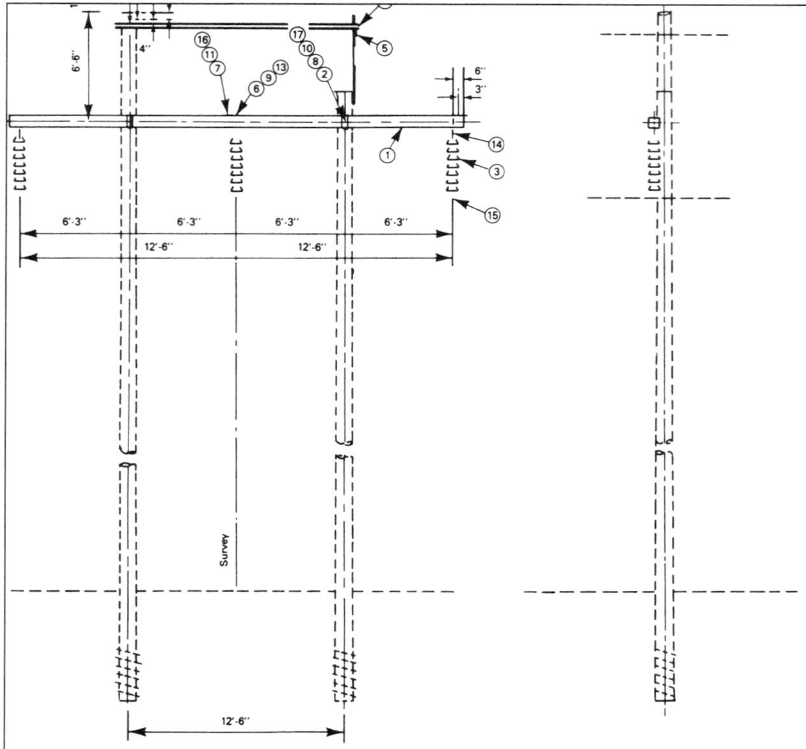

DRG. REF.	REQ'D	DESCRIPTION	ITEM
1	1	5 5/8" x 7 5/8" x 26'-0" Wood Crossarm, Type 55	g
2	2	Reinforcing Plate for 8" Crossarm	eg
3	*	5 3/4" x 10" Suspension Insulator	k
4	2	Ground Wire Cable Support	ed
5	2	Ground Wire Suspension Clamp	m
6	3	3/4" x 10" Eye Bolt	o
7	3	1/2" x 8" Machine Bolt	c
8	2	7/8" x reqd length, Machine Bolt	d
9	10	4" x 4" x 3/16" Galv. Sq. Washer, 13/16" Hole	d
10	2	Locknuts for 7/8" Bolt	ek
11	3	Locknuts for 1/2" Bolt	ek
12	4	Locknuts for 5/8" Bolt	ek
13	5	Locknuts for 3/4" Bolt	ek
14	3	Suspension Hook	eh
15	3	Suspension Clamp and Connecting Piece	ei
16	6	2 1/4" x 2 1/4" galv. Sq. Washer, 9/16" Hole	
17	2	4" x 4" x 3/16" Sq. Curved Washer, 15/16" Hole	
		Steel Bayonet	
		Steel Tie Bar	

LIST OF MATERIAL

X-BRACED
TRANSMISSION LINE TANGENT STRUCTURE
H-FRAME SUSPENSION - TWO POLE
(115 KV. MAXIMUM)

Scale: None		Date: ______
		TH-I AM

*As required. See Drg. TM-1

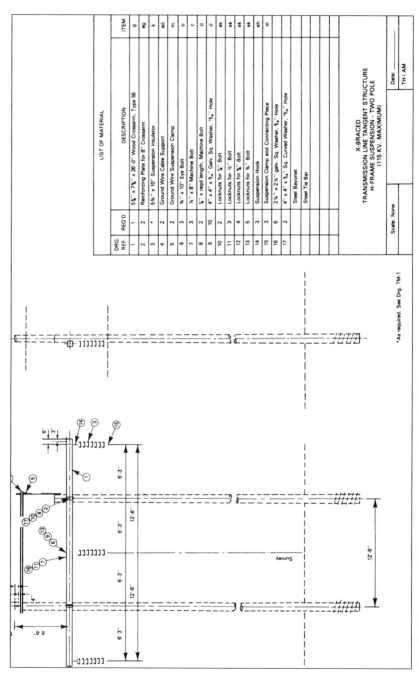

Fig. B–2 Two-pole, H-frame transmission line, TH-1AM specifications

The original 69-kV line had a design ruling span of 500 ft as compared with an actual ruling span of 380 ft based on structure locations. A check of several contemporary line projects revealed a similar practice.

Summation of moments at the groundline because of wind and ice loads revealed that the larger 477 mcm 26/7 conductor and two 3/8" H.S. steel shield wires could be installed on the uprated TH-1AM tangent structure. This would provide a 4.0 safety factor under NESC heavy-loading criteria. The maximum allowable sums of adjacent spans for several typical structure heights are given in Table B–2.

Table B–2 Maximum Sums of Adjacent Spans, TH-1AM Structures of Several Heights

| TH-1AM Structure | | Maximum Sum of Adjacent |
Height (ft)	Class	Spans (Safety factor = 4.0)
50	3	1,151 ft
55	3	1,136 ft
60	3	1,129 ft

The single cross-arm (Type 55) of the TH-1AM structure has a maximum vertical span limitation of 822 ft under NESC heavy-loading conditions and a 4.0 safety factor.

The allowable sum of adjacent spans and the vertical span allowed by the cross-arm strength permitted existing poles and pole locations to be used in spotting all tangent structures in the uprated line. Additional span lengths could have been attained by installing X-braces or using an assembled cross-arm. The span lengths in the existing line, however, allowed the use of the unbraced, single-piece cross-arm structure as described.

As a general rule, standard REA angle and dead-end structures were called for at line angles and dead-end points in the uprated line. No attempt was made to redesign or uprate existing 69-kV structure configurations. This rule was employed to assure ample strength and electrical clearances at these critical line locations.

The galloping ellipse patterns were calculated and plotted for the longest individual span in the line (590 ft). The single loop method of analysis was used and less than 10% overlap was detected. Therefore, the structure spacing, wire sags, and maximum span length were concluded to be satisfactory.

Inspection/Survey

A vital part of the design procedure in a line uprating is a thorough inspection of the existing line. A competent inspection crew should be assigned to foot patrol and inspect the entire line length. The inspection crew should use copies of the existing line plan and profile drawings. They should perform many checks including the following:

- Verify the elevation of the elevation profile
- Verify the height and groundline circumference of each existing structure
- Verify the height and survey station of each utility crossing
- Report the location and extent of any right-of-way encroachment (house, barn, mobile home, radio/TV antenna, etc.)
- Verify survey station of all highway/road/railroad crossings
- Report any land-use changes (original pastureland now under cultivation, pivot irrigation now installed, etc.)
- Report any terrain changes (excavations, landfills, stock tanks, terraces, etc.)
- Report any structures requiring line maintenance (ground or internal rot, damaged pole, etc.)

Line conversion or voltage uprating may warrant a complete new line survey. If the existing transit and level data are relatively recent (less than 10 years old) and the line route is not in a developing urban area, the existing survey information may be adequate. It must, however, be verified thoroughly by the foot patrol inspection noted above.

An important point to recognize in line uprating is the fact that the centerline of the uprated structure may not be the same centerline of the original structure. The original centerline of the TS-1 structure coincided with the original survey centerline. However, the centerline of the uprated TH-1AM structure was offset 1.91 m (6 ft 3 in.) from the original survey centerline. Because of the flatness of the terrain passed through in this study, the offset did not create a real problem. However, a similar offset through heavily wooded or side-sloping terrain might have been prohibitive.

Right-of-Way

The original TS-1 single pole 69-kV transmission line required only 15.24 m (50 ft) of right-of-way—7.62 m (25 ft) either side of the centerline. The uprated TH-1AM two-pole 115-kV line requires a minimum of 22.8 m (75 ft) of permanent right-of-way. This is based on 9.14 m (30 ft) from one side to the original centerline plus 12.19 m (40 ft) on the other side to allow for centerline and conductor offset, blow-out, and electrical clearance.

Notification

A "notice of intent to construct" is usually required by local regulatory agencies. This type of notification should be given even if it is not required by statute. Information concerning line route, structure configuration, wire size, minimum ground clearances, operating voltage, etc. should be sent to appropriate agencies. These include local utilities, highway departments, county officials, pipeline companies, rural water cooperatives, the Federal Aviation Administration, or other groups that may have facilities in the area or be affected by the new construction.

Design Data: TH-1AM Uprated Line

Design summary data for the TH-1AM uprated line and structure are given in Figures B–3 , B–4, B–5 , and B–6, and pages following. One of the main advantages of using the TH-1AM configuration is that it allows an increased ground clearance of 1.68 m (5 ft 6 in.) over the ground clearance allowed by the original TS-1 structure. This provided enough additional ground clearance to allow use of larger conductors as well as voltage upgrading. An additional 0.46 m (1 ft 6 in.) of ground clearance could be provided by raising the cross-arm to within 0.3048 m (1 ft 0 in.) of the pole top of the original structure.

The lightning shield angle of the TH-1AM structure is approximately 28° to the outside phases.

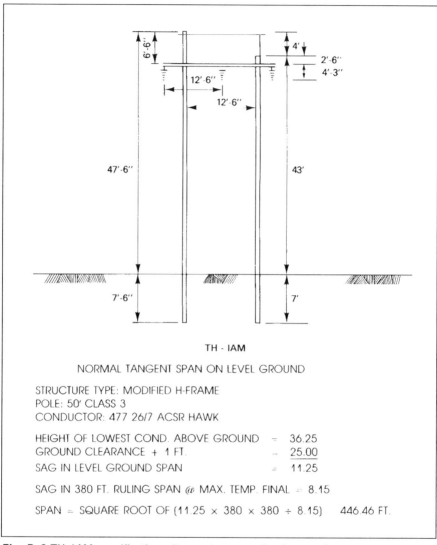

TH - IAM

NORMAL TANGENT SPAN ON LEVEL GROUND

STRUCTURE TYPE: MODIFIED H-FRAME
POLE: 50' CLASS 3
CONDUCTOR: 477 26/7 ACSR HAWK

HEIGHT OF LOWEST COND. ABOVE GROUND = 36.25
GROUND CLEARANCE + 1 FT.   -- 25.00
SAG IN LEVEL GROUND SPAN   = 11.25

SAG IN 380 FT. RULING SPAN @ MAX. TEMP. FINAL = 8.15

SPAN = SQUARE ROOT OF (11.25 × 380 × 380 ÷ 8.15)   446.46 FT.

Fig. B–3 TH-1AM specifications (tangent span on level ground)

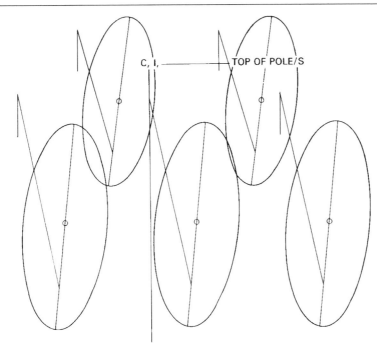

FOR SPAN LENGTH OF 590 FEET
SINGLE LOOP   LISSAJOUS   ELLIPSE PATTERN
STRUCTURE: MODIFIED H-FRAME

DATA TABLE FOR WIRE CODEWORD ⅜" H.S. STEEL

ANGLE   15.672537 AND ANGLE + ANGLE/2 = 23.508806 DEGREES
A + E = 12.197964 FEET
MAGNITUDE OF MAJOR AXIS   16.247455 FEET
MAGNITUDE OF MINOR AXIS   6.498982 FEET
B = 3.049491 FEET

POINT OF ATTACHMENT = −3.25   DOWN AND 6.75 FROM CENTER LINE
POINT OF ATTACHMENT   −3.25   DOWN AND −6.75 FROM CENTER LINE

DATA TABLE FOR WIRE CODEWORD HAWK

ANGLE   11.652755 AND ANGLE + ANGLE/2 = 17.479133 DEGREES
A + E = 18.851455 FEET
MAGNITUDE OF MAJOR AXIS = 19.652519 FEET
MAGNITUDE OF MINOR AXIS   7.861008 FEET
B = 3.730504 FEET

POINT OF ATTACHMENT = 2.75   DOWN AND 12.5 FROM CENTER LINE
POINT OF ATTACHMENT = 2.75   DOWN AND 0 FROM CENTER LINE
POINT OF ATTACHMENT = 2.75   DOWN AND −12.5 FROM CENTER LINE

Fig. B–4 TH-1AM wire specifications

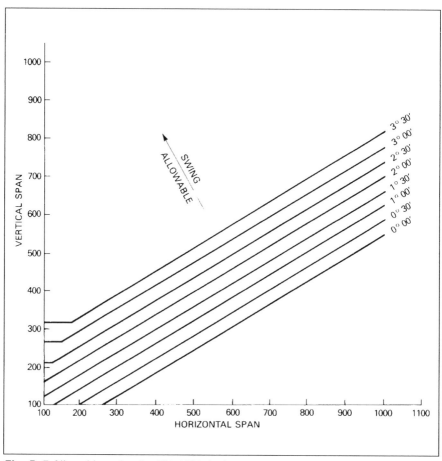

Fig. B–5 Allowable swing for TH-1AM design

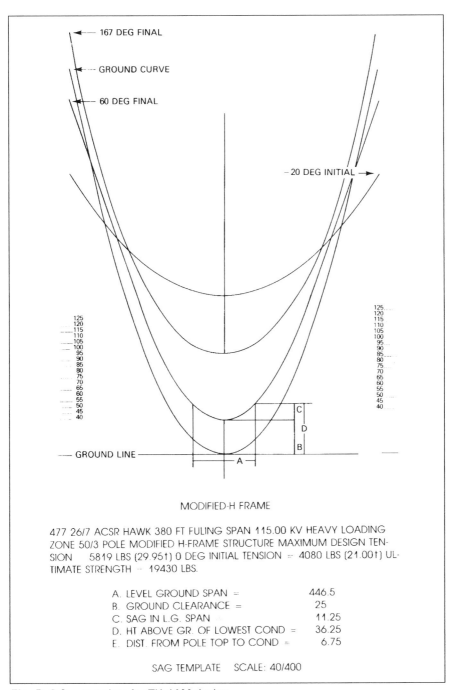

167 DEG FINAL

GROUND CURVE

60 DEG FINAL

20 DEG INITIAL

125
120
115
110
105
100
95
90
85
80
75
70
65
60
55
50
45
40

125
120
115
110
105
100
95
90
85
80
75
70
65
60
55
50
45
40

C
D
B

GROUND LINE

A

MODIFIED-H FRAME

477 26/7 ACSR HAWK 380 FT FULING SPAN 115.00 KV HEAVY LOADING ZONE 50/3 POLE MODIFIED H-FRAME STRUCTURE MAXIMUM DESIGN TENSION   5819 LBS (29.95↑) 0 DEG INITIAL TENSION = 4080 LBS (21.00↑) ULTIMATE STRENGTH = 19430 LBS.

A. LEVEL GROUND SPAN =	446.5	
B. GROUND CLEARANCE =	25	
C. SAG IN L.G. SPAN	11.25	
D. HT ABOVE GR. OF LOWEST COND =	36.25	
E. DIST. FROM POLE TOP TO COND =	6.75	

SAG TEMPLATE   SCALE: 40/400

Fig. B–6 Sag template for TH-1AM design

Appendix

Basic Electricity

Basic Electric Circuit

The basic circuit for transmitting electrical energy must consist of two paths or conductors, one sending and the other a return, together forming a continuous path or a closed circuit for the electric current to flow. This holds for both DC and AC systems.

The conductors or wires in an electric circuit can be thought of as electrical pipes through which flow a stream of electrons, constituting a flow of electricity (refer to chap. 1 for a water analogy).

Electric Pressure or Voltage

For the electrons to stream through the wires, an electrical pressure is necessary. This pressure can be created by a generator that may be likened to a pump. The electrical pressure is expressed in volts (named in honor of the Italian physicist Alessandro Volta).

Current or Amperage

Electric current is the rate of flow of electrons in a circuit. Similar to the flow of water measured in gallons or litres per second, the number of electrons passing a reference point in 1 second determines the current strength. This is expressed in amperes (named after the French physicist André-Marie Ampère). Scientific measurement has determined that 6.29 billion electrons passing in 1 second make up 1 ampere.

Resistance

Just as water flowing in a pipe is resisted by friction between the water and the pipe, electrons encounter resistance to their flow in conductors. A fixed volume of water encounters less resistance to its flow in a large diameter pipe than in one of small diameter. This is also true for electric current; a fixed value of electric current will encounter less resistance in a large diameter conductor than in one of small diameter. The electrical resistance is expressed in ohms (after Georg Simon Ohm, the German physicist).

Ohm's Law

In any electrical circuit, the three factors of pressure, current, and resistance are interrelated.

In a circuit, the current flow will vary directly with the pressure applied, and inversely with the resistance of the circuit. This can be expressed as:

$$\text{Current in amperes} = \frac{\text{Pressure in volts}}{\text{Resistance in ohms}}$$

This is known as Ohm's Law. Expressed with variables, the law becomes:

$$I = \frac{E}{R}$$

where

I	is the intensity of the current in amperes
E	is the pressure in volts
R	is the resistance in ohms

If any two quantities are known, the third may be found by applying this equation.

Types of Circuits

There are two basic types of electric circuits: the series circuit and the multiple or parallel circuit; other types are a combination of the two.

Series Circuit

In a series circuit, all the parts making up the circuit are connected in succession, so that the current through all of the parts is the same.

The circuit shown in Figure C-1 contains four resistances connected in series with the same current through each of them. Assume an electrical pressure or voltage of 120 volts (V) is applied across the terminals of the circuit and a current of 6 amperes (A) flows through the circuit. Then, by Ohm's Law, the total resistance of the circuit will be found to be 20 ohms (Ω); that is:

$$I = \frac{E}{R} \text{ or } 6 = \frac{120}{R}; \text{ solving for } R = \frac{120}{6} = 20 \text{ ohms}$$

Voltages measured across each of the four resistances (R_1, R_2, R_3, and R_4) shown are found to be:

$$E_1 = 18 \text{ V}; E_2 = 30 \text{ V}; E_3 = 48 \text{ V}; E_4 = 24 \text{ V}$$

Applying Ohm's Law to each part of the circuit, the resistance values can be found:

$$I = \frac{E}{R} \text{ from which } R = \frac{E}{I}$$

Hence

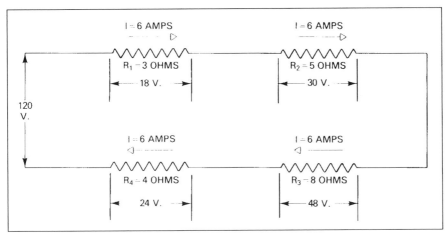

Fig. C-1 Simple series circuit

$$R_1 = \frac{18}{6} = 3 \text{ ohms} \quad R_2 = \frac{30}{6} = 5 \text{ ohms}$$

$$R_3 = \frac{48}{6} = 8 \text{ ohms} \quad R_4 = \frac{24}{6} = 4 \text{ ohms}$$

We can check these values. The sum of the four voltages across each of the resistance (the drop in pressure as the current flows in the resistance) will be:

$$E_1 + E_2 + E_3 + E_4 = E$$

or

$$18 \text{ V} + 30 \text{ V} + 48 \text{ V} + 24 \text{ V} = 120 \text{ V}$$

The sum of the separate resistances is equal to the total resistance:

$$R_1 + R_2 + R_3 + R_4 = R$$

or

$$3 \text{ ohms} + 5 \text{ ohms} + 8 \text{ ohms} + 4 \text{ ohms} = 20 \text{ ohms}$$

Note that there is a drop in voltage through each resistance. From Ohm's Law, this drop will be the product of the current and resistance, that is $E = IR$, and is usually referred to as the IR drop. Note also that the voltage drop in each part is proportional to its resistance and that the sum of all the voltage drops is equal to the applied voltage. (Resistance of the connecting wires has been neglected for purposes of illustration.)

Multiple or Parallel Circuits

In a multiple or parallel circuit, all of the components are connected so as to receive full line voltage, and the current that flows in each component depends on its resistance.

If the same four resistances are connected in parallel across the same line voltage, as shown in Figure C-2, applying Ohm's Law, the currents flowing in each resistance have the following values:

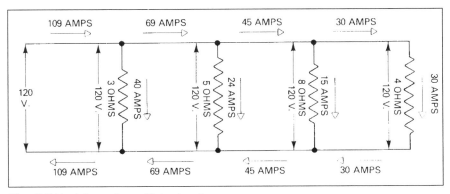

Fig. C-2 Resistance in multiple circuits

Since

$$R_1 = 3 \text{ ohms}, I_1 = \frac{E}{R_1} = \frac{120}{3} = 40 \text{ amperes}$$

$$R_2 = 5 \text{ ohms}, I_2 = \frac{E}{R_2} = \frac{120}{5} = 24 \text{ amperes}$$

$$R_3 = 8 \text{ ohms}, I_3 = \frac{E}{R_3} = \frac{120}{8} = 15 \text{ amperes}$$

$$R_4 = 4 \text{ ohms}, I_4 = \frac{E}{R_4} = \frac{120}{4} = 30 \text{ amperes}$$

The total current will be:

$$I_1 + I_2 + I_3 + I_4 = I$$

or

40 amperes + 24 amperes + 15 amperes + 30 amperes = 109 amperes

The resistance of the entire circuit, applying Ohm's Law, will be:

$$R = \frac{E}{I} = \frac{120}{109} = 1.101 \text{ ohms}$$

Another way of obtaining the same result is to add the reciprocals of each resistance and taking the reciprocal of the sum of the reciprocals.

$$\frac{1}{R_1} + \frac{1}{R_2} + \frac{1}{R_3} + \frac{1}{R_4} = R$$

$$\frac{1}{3} + \frac{1}{5} + \frac{1}{8} + \frac{1}{4} = \frac{1}{R}$$

$$0.333 + 0.200 + 0.125 + 0.250 = 0.908$$

$$R = \frac{1}{0.908} = 1.101 \text{ ohms}$$

Note that the resultant resistance is always less than the smallest of the component resistances. Each additional resistance connected in parallel adds an additional path for the current to flow; as the conducting paths are increased, the total resistance is lowered.

Series-Parallel Circuit

An example of resistances connected in series-parallel is shown in Figure C-3. To obtain the total resistance of this circuit, the resultant resistance of each of the parallel groups is first determined, then added to the resistances in series. The same process applies to any number and kind of groups of resistances.

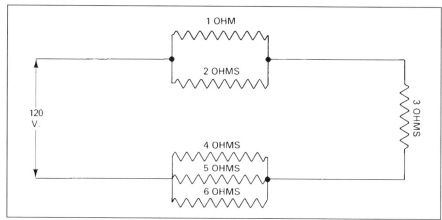

Fig. C-3 Series multiple circuits

Exercise C.1: Find the voltages across each of the resistances, the currents flowing in them, and the total resistance of the circuit.

Power

When electricity flows through a resistance, electrical energy is consumed in that resistance. The *rate* of consumption of electrical energy is known as electrical power. The unit of electrical power is called the watt (W) (after James Watt, the developer of the steam engine). The watt is too small for practical purposes, and the kilowatt (kW—equal to 1,000 W) is most frequently used.

Power is the *rate* of doing work. Electrically, this depends on the electrical pressure applied and the current flowing in a device, represented as a resistance. Hence:

$$\text{Power (W)} = \text{Pressure (V)} \times \text{Current (A)}$$

By experiment, it is found that 746 W of electrical power are equivalent to 1 horsepower (hp) of mechanical power.

Energy

Since power is the rate of expending energy, then energy expended will be the product of power and the time it is applied:

$$\text{Energy} = \text{Power} \times \text{Time}$$

Time may be expressed in any unit: seconds, minutes, hours, days, etc. The common unit is the hour (hr); hence:

$$\text{Energy} = \text{Power (kW)} \times \text{Time (hr)}$$

The most commonly used unit of electrical energy is the kilowatt-hour (kWh).

Heat Loss

When power flowing in a circuit does not produce useful work, that is, where the electrical energy is not converted to some mechanical work, it is converted into heat. This heat is dissipated into the surrounding atmosphere. Owing to the electrical resistance encountered, it may be likened to the heat developed by friction, and represents a loss. This may be determined by applying Ohm's Law:

If:

$$Power (W) = Voltage \times Current, or W = E \times I$$

and

$$Voltage = Current \times Resistance, or E = IR$$

then

$$Power (W) = Voltage (IR) \times Current (I)$$
$$W = IR \times I$$
$$W = I^2 R$$

This represents the rate of energy loss given off as heat; actual energy loss multiplies this value by the time it is expended and is measured in kilowatt-hours.

Note that if the current (I) flowing in a resistance (conductor) is doubled, the heat loss is not doubled, but quadrupled.

In transmitting power over great distances, two factors must be considered:

1. The voltage (IR) drop in the line must not be so great that insufficient electrical pressure or voltage will result at the receiving end.
2. The power (I^2R) loss in the line must not be so great that other means of supplying power may prove more economical. (One example would be to set up a generating plant at or near the receiving end and transport fuel there.)

Inductance

There is a basic relationship between electricity and magnetism. An electric current flowing in a conductor will produce a magnetic field around it. A conductor cutting the magnetic field will have an electric voltage induced in it, and a current will flow if the conductor is part of a circuit. In the latter case, the magnitude of the voltage produced depends on the length of the conductor cutting the field. It also depends on the speed at which the conductor cuts the magnetic field, and the density of the magnetic field.

An AC-carrying conductor will have a magnetic field around it that alternates its characteristics as the current in the conductor alternates with its frequency. The magnetic field builds up to a maximum in one direction, reduces to 0, builds up to a maximum in the opposite direction, and again reduces to 0, completing one cycle. (Frequency is the number of such alternations, or cycles, occurring in 1 second.)

Self-Inductance

The alternating magnetic field around the conductor will cut the conductor, inducing in it a voltage distinct from that causing the original current to flow. This second voltage will also produce a current that, in turn, affects the original current. This, in turn, affects the magnetic field around the conductor, thus affecting the entire relationship that finally stabilizes at some point.

The currents flowing in the conductor, the original and induced, are not in step, or "in phase." In other words, the rising and falling of their cycles do not coincide. Their relationship may be shown in Figure C-4; the two voltage waves will be displaced a quarter cycle from each other. Actually, of course, only the one resultant voltage exists in the conductor.

The current values, however, still reach their maximum and 0 values as they did originally. The net effect of the reaction in the conductor is to cause the current to "lag" behind the voltage, as illustrated in Figure C-4c. It will be noted then that the current and voltages do not act together throughout the cycle. Only some portion of the current will act in conjunction with the voltage. The power produced then will be the product of the voltage and current values at any particular point and not the product of the voltage and current when acting together, or "in phase." The ratio of the first quantity to the latter is known as the "power factor."

Mutual Inductance

This same reaction may be caused by the magnetic fields of adjacent conductors and, since both conductors affect each other, it is called "mutual inductance" (see Fig. C-5).

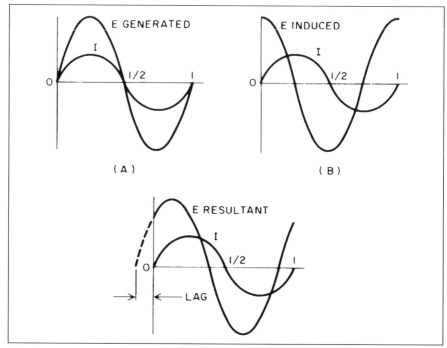

Fig. C-4 Effect of inductance on voltage and current in a conductor (not to scale)

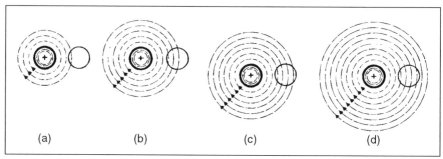

Fig. C-5 Effect of the magnetic field about a conductor on an adjacent conductor

Inductive Reactance

The effect of this inductance is to act as an obstruction to the flow of current in a circuit, similar to, but greatly different than, the effect of resistance. Such an obstruction is known as the "reactance" of the circuit and is also expressed in ohms. (To avoid confusion, the letter R is used to denote resistance and X_L to denote reactance due to inductance.)

Ohm's Law can also be used to find the current flow in an inductive circuit, assuming the circuit has no resistance. Hence:

$$I = \frac{E}{X_L}$$

Note, however, that when only inductance obstructs the flow of electricity, the current wave "lags" the voltage wave by a quarter cycle (see Fig. C-4b).

Resistance and Inductance

We can apply Ohm's Law to each of two circuits, one containing resistance only, the other inductive reactance only. Assuming a voltage of 120 V is applied to each, presenting an obstruction equivalent to 2 Ω, then:

$$I_1 = \frac{E}{R} = \frac{120}{2} = 60 \text{ amperes} \quad I_2 = \frac{E}{X_L} = \frac{120}{2} = 60 \text{ amperes}$$

Now assume a circuit that has a combination of both resistance and inductive reactance, with the same values indicated above. With a voltage of 120 V applied, the current will be found to be 42.6 A. By Ohm's Law:

$$R = \frac{120}{42.6} = 2.82 \text{ ohms}$$

The resultant obstruction to the flow of electricity is greater than either the resistance or the inductive reactance (each 2 Ω) but is *not* equal to the arithmetic sum (4 Ω). By analysis, it will be found that the resultant obstruction may be found by obtaining the square root of the sum of the squared values of resistance and inductive reactance:

$$Z = \sqrt{R^2 + X_L^2} \quad \text{or} \quad \sqrt{2^2 + 2^2} = \sqrt{8} = 2.83 \text{ ohms}$$

where

Z   is the resultant obstruction (referred to as "impedance" to differentiate it from its component resistance and inductive reactance)

From Figure C-6b it will be noted that when resistance is combined

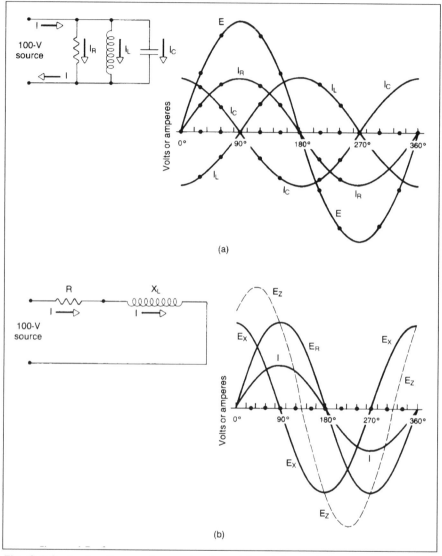

Fig. C-6 Cycle of voltage and current values: (**a**) resistance, inductance, and capacitance in parallel; (**b**) resistance and inductance in series

with inductance, the current lags the voltage wave less than when the circuit contained only inductance. As more resistance is added, the current and voltage waves will closely approach each other.

Capacitance

There is another reaction that occurs in AC-carrying conductors that is not due to the magnetic fields around them. This is an electrostatic effect that takes place between conductors. During one-half cycle of the alternation, there will be a scarcity of electrons in a conductor (with reference to a fixed point), and it will tend to attract electrons from an adjacent conductor. During the next half cycle, a reverse action takes place—an excess of electrons in the conductor, and a tendency for electrons to flow to the adjacent conductor. Thus a to-and-fro, or alternating, circulation of electrons is set up in the second conductor; that is, an AC is set up in the second conductor.

If both conductors carry AC, they will react on each other as described above. The amount of this reaction is called "capacitance" and will depend on the areas of the conductors exposed to each other. The distance and kind of insulating material between them and the number of electrons involved depend on the voltage or current in the conductor.

Similar to inductance, capacitance will set up a distinct current in the conductor that will be displaced by a quarter cycle from the normal line current. In this case, however, the current will "lead" the voltage wave by a quarter cycle (see Fig. C-7).

Such obstruction of the capacitance or capacitor to the flow of electricity in a conductor is referred to as capacitive reactance and is also measured in ohms. Capacitive reactance is denoted by the letters X_C to distinguish it from other quantities.

Water Analogy of a Capacitor

A water analogy can be used in the discussion of capacitors (see Fig. C-8). As the pump moves to the right, water flows in the top pipe to the cylinder at the right. The flexible diaphragm in the cylinder will move to the right (as shown), transmitting its motion to the water to its right during the motion of the pump. In the leftward motion of the pump, the reverse action takes place. The net result is that an alternating flow of water is set up in both the pipes, without a direct connection between the water in the lower pipe and that in the upper pipe.

In the process, because of the flexibility of the diaphragm, the current of water will tend to arrive at the undistended diaphragm before the pump completes its stroke in one direction. This may be observed by the relative positions of the driving motor on the left and the water in the pipes driving a

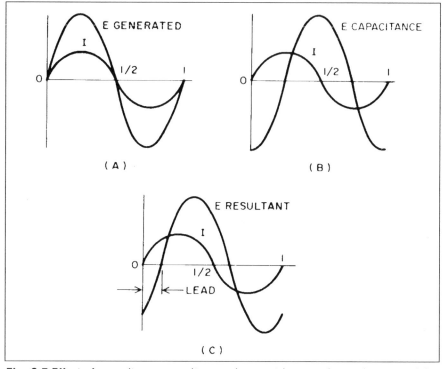

Fig. C-7 Effect of capacitance on voltage and current in a conductor (not to scale)

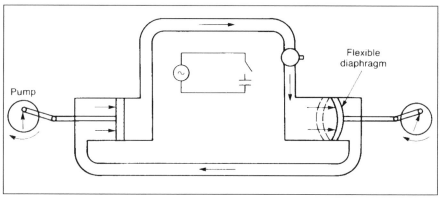

Fig. C-8 Water analogy of a capacitor in an AC circuit

water motor. The current flow may be said to "lead," or act ahead of, the pressure being applied by the pump at the left. The mechanical stresses on the diaphragm, alternately compression on one side and elongation on the other, correspond to the electrostatic stresses on the dielectric in the capacitor.

Resistance and Capacitance

The same phenomenon as existed in an inductive circuit exists in the capacitive circuit, except that the obstructive effects are in opposition (180° apart). In a circuit containing both resistance and capacitance, the resulting obstruction is obtained in the same manner as for resistance and inductance. That is:

$$Z = \sqrt{R^2 + X_C^2}$$

In this case, the more resistance added, the closer the voltage and current waves will approach each other.

Impedance

The total obstruction to the flow of current in a circuit may be caused by resistance, inductance, and capacitance. The effects of inductance and capacitance are generally referred to as reactance. Hence, the combined effect of resistance and reactance is called the "impedance" of the circuit and, as already noted, is designated by the letter Z. For AC circuits, therefore, Ohm's Law becomes:

$$I = \frac{E}{Z}$$

Resistance, Inductance, and Capacitance

As mentioned earlier, the effects of inductance and capacitance are in direct opposition to each other. Hence, the net reactance will be the difference between the inductive reactance and capacitive reactance, that is:

$$X = X_L - X_C$$

The resultant impedance of a circuit containing resistance and reactance will then be:

$$Z = \sqrt{R^2 + X^2}$$

Or substituting:

$$Z = \sqrt{R^2 + (X_L - X_C)^2}$$

The current flowing in the circuit will be calculated by Ohm's Law:

$$I = \frac{E}{Z} = \frac{E}{R^2 + (X_L - X_C)^2}$$

Resonance

It is evident that if $X_L = X_C$, the only quantity left will be R, the resistance. When this condition occurs, the circuit is said to be in "resonance."

The relative value of resistance, inductance, and capacitance of a circuit will determine the relative position of the current wave with respect to the voltage wave.

Power in AC Circuits

As mentioned earlier, power is the rate at which electrical energy is transformed into heat or mechanical energy and is equal to the product of the voltage and current. In an AC circuit, the change of energy at any moment is the product of voltage and current at that instant. Power values for circuits having only resistance, only inductance, and only capacitance are shown in Figure C-9. Note that real power is produced only in the resistance circuit. In the inductance and capacitance circuits, the net result of the power curves is 0. In other words, the areas under the curves above the axis are exactly equal to those below the axis.

In both the inductance and capacitance circuits, electric energy is stored in the magnetic and electrostatic fields during the time the current is increasing. The energy is returned back to the circuits when the current is decreasing.

Power Factor

As indicated earlier, the product of voltage and current does not indicate the true power flow in the circuit. Such a product is termed "apparent

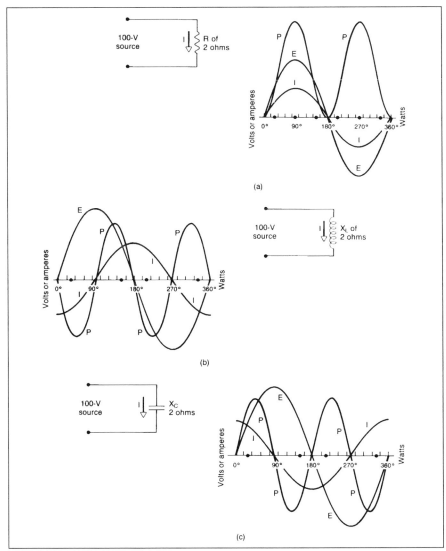

Fig. C-9 Power values in (**a**) resistance; (**b**) inductance; and (**c**) capacitive circuits

power" and is expressed in volt-amperes, or kilovolt-amperes (kVA), instead of watts or kilowatts (kW). To calculate the true or real power, the apparent power is multiplied by a factor called the "power factor."

The power factor of a circuit is the ratio between the true and the apparent power and is usually expressed as a decimal or percent:

$$\text{Power Factor} = \frac{\text{True power}}{\text{Apparent power}} = \frac{\text{Watts}}{\text{Volt-amperes}} = \frac{\text{kW}}{\text{kVA}}$$

In the resistance circuit, all of the energy delivered to it is converted into heat or mechanical energy. Consequently, the true power and the apparent power are the same, and the power factor is 1 (or 100%). In the inductance or capacitance circuit, the true power is 0, while the apparent power may have a value. The power factor for both, however, is 0, or 0%.

In a circuit containing resistance, inductance, and capacitance, the real power and apparent power are unequal. Thus the power factor will have a value somewhere between 0 and 100%, depending on the relative values of the resistance, inductance, and capacitance components.

Effective Values of Voltage and Current

As mentioned earlier, when an electric current flows through a circuit, heat is generated or produced in the circuit. The rate at which heat is produced in the circuit is equal to I^2R; where I is the current in amperes and R is the resistance in ohms. This heating effect takes place when either AC or DC flows.

In an AC circuit, the rate at which heat is being generated is constantly changing because the current is constantly changing. In a DC circuit, when a current flows, the rate at which heat is developed is constant (for a fixed value of current) because the direct current remains constant in value. For example, the alternating current that reaches a maximum value of 100 A would not have the same heating effect as a direct current of 100 A. To produce the same heating effect, the alternating current would have to be of such value that it reaches a maximum value of 141.4 A. Such a current varying according to a sine wave between 0 A and a maximum of 141.4 A has the same effective heating value as a direct (constant) current of 100 A. Hence, its "effective" value is said to be 100 A.

The same effective value is used for voltages. Thus an alternating (sine wave) voltage varying between 0 V and 141.4 V is said to have an "effective" value of 100 V.

Both alternating currents and voltages are usually expressed in terms of effective values. The ratio between the effective and maximum values is illustrated in Figure C-10. This ratio can be expressed in either of two ways:

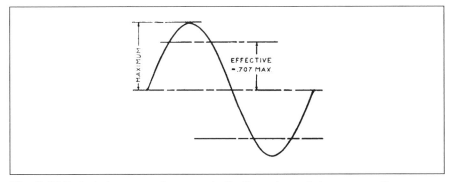

Fig. C-10 Ratio between maximum and effective values of sine wave voltage or current

$$\frac{\text{Effective value}}{\text{Maximum value}} = \frac{0.707 \times \text{Maximum value}}{1.414 \times \text{Effective value}}$$

Thus, if one value is known, the other can be easily determined.

The maximum value is an instantaneous value. Any other instantaneous values for other instants during the cycle can be determined by scaling off values from the curve of a sine wave.

AC instruments, such as ammeters and voltmeters, are calibrated to indicate effective values.

Basic Electric Circuit

The basic circuit for transmitting electrical energy consists of two paths or conductors, one sending and the other returning. This holds for both AC and DC systems. In DC systems, the voltage applied is continuous, and the current flows in one direction continuously. In AC systems the voltage applied rises from 0 to a maximum and drops back to 0. It then reverses direction and goes through the same variations. A graph of this pattern is called a "sine wave" (see Fig. C-11). Current flows back and forth in a push-pull fashion with each half of the cycle doing productive work.

In the United States, systems operate at 60 cycles per second, that is, there are 60 of these push-pull flows each second. In other parts of the world, many systems operate at 50 cycles per second.

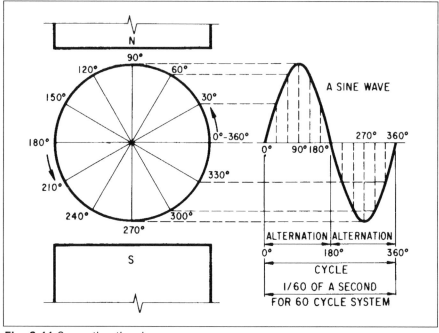

Fig. C-11 Generating the sine wave

Ground Neutral

One such two-wire circuit (see Fig. C-12), whether AC or DC, can carry a certain prescribed amount of power. For safety reasons, usually one conductor—the return—is grounded, that is, connected to earth. The voltage applied, then, can be specified as so many volts above or from ground. If more transmitted power is desired, a second similar circuit can be added.

In the latter circuit, the voltage applied can be in the reverse direction of the first, that is, so many volts below ground. If the first voltage applied is labeled positive, then the second can be labeled negative. It will be noted that both circuits have one conductor grounded; hence, these two ground, or return, conductors can be connected together. It will further be observed that the current in one return circuit opposes that of the other circuit. The common return—or neutral—conductor will now carry only the difference in magnitude of these two currents. The neutral conductor, therefore, need not be as large as either of the main conductors, resulting in economy.

Further, if both circuits carried exactly equal currents, the current in the neutral would be 0, and the neutral conductor could be eliminated,

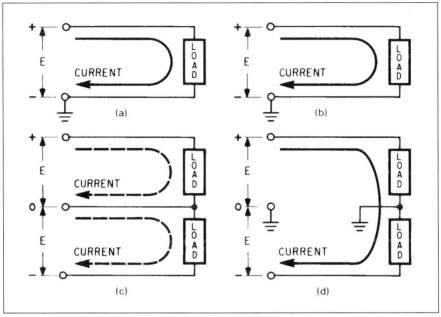

Fig. C-12 Effects of ground/neutral connections. (**a**) a schematic two-wire DC transmission line where *E* is voltage from the line to the supporting structure; (**b**) schematic DC transmission line using earth return; (**c**) three-wire DC system where current is cancelled in the center leg; (**d**) schematic ± *E* DC grounded transmission line with twice the current flow capacity of (**a**) and same insulation and conductor size

resulting in further economy. The voltage difference between both main conductors will now be double that for one. However, since each conductor carries insulation for its original voltage, the insulations existing between both conductors will be sufficient to sustain the double voltage.

Polyphase Systems

For AC systems, the power supplied can be delivered over more than two circuits, as described. These are known as polyphase (or multiphase) circuits. In a single-phase circuit, only one phase or set of voltages of sine-wave form is applied to the circuit and only one phase of sine-wave current flows in the circuit. In polyphase circuits, two or more phases or sets of sine-wave voltages are applied to the different portions of the circuits. A corresponding number of sine-wave currents will flow in those portions of the circuits.

The different portions of the polyphase circuits are usually called the "phase." Studies indicate an economic and practical arrangement is to limit this number to three, and this is referred to as a three-phase circuit. Practically all AC transmission systems are three phase and have three current-carrying conductors. The phases are usually lettered to identify them, as the "A" phase, "B" phase, and "C" phase. The voltages applied to the separate phases of the circuit are correspondingly referred to as the "A-phase voltage," the "B-phase voltage," etc. The phase currents are similarly identified.

In a three-phase circuit, the alternating pulses of electricity are displaced one from another by 120°, that is, one complete cycle is considered to represent 360°. Then, if the voltage at one conductor (A) starts at zero, the voltage in a second conductor (B) with reference to the first will begin its cycle 120° later. That in the third conductor (C), with reference to the first, will begin 240° later (see Fig. C-13). The fundamental principles of the flow of AC are the same whether applied to single-phase or polyphase circuits.

The voltages for polyphase systems are supplied from polyphase (multiphase) generators, each phase of voltage generated in a separate coil (or coils connected in parallel). The separate coils are arranged for connection in different ways to form the polyphase system. Two commonly used methods of connecting the coils of three-phase generators to supply three-phase transmission systems are shown in Figure C-14. One method employs the "delta" connection, the other the "wye" or "star" connection.

Delta and Wye Connections

The voltages between the terminals or conductors of the two types of three-phase systems are shown in Figure C-14. Note the voltages between the terminals are the same for both systems. In the delta system, the full (terminal) voltage is imposed on the coils of the generator or motors whose coils are also connected in delta and that may be connected to the delta system. In the wye system, the voltage imposed on these coils is only 0.866 that of the delta system.

The common connection of the three phases in the wye system may be connected to a neutral or fourth conductor, or may be left as is to "float." The shield wire may also be the fourth wire and carries the unbalance of one or more such three-phase circuits. In some cases, it is also connected to or serves as the neutral conductor for distribution circuits that may be installed on the same supporting structures.

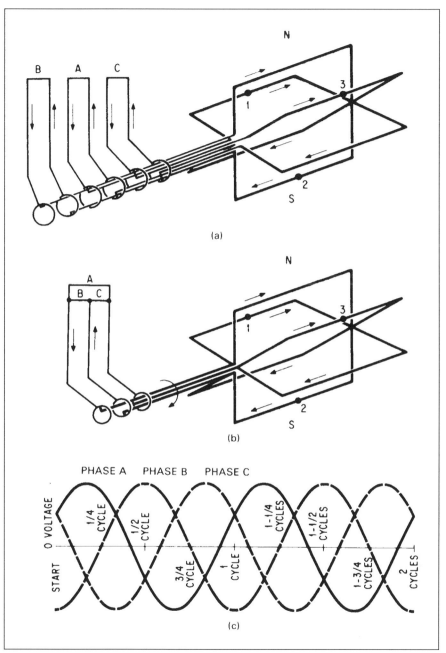

(a)

(b)

(c)

Fig. C-13 Three-phase generator: (**a**) three single-phase generators mounted on the same shaft; (**b**) a three-phase generator; (**c**) curves showing voltages in a three-phase generator

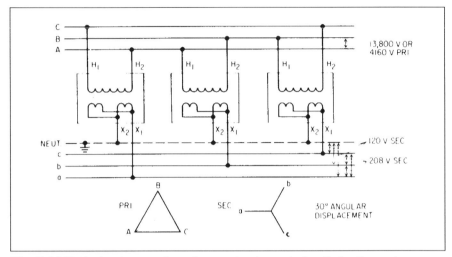

Fig. C-14 Methods of connection of generator (or motor) coils for three-phase systems

The coils connected in the delta system are subjected to higher voltages than coils in the wye system. As there is no ground connection, one or more grounds (faults) on only one phase will not affect the operation of the system. A second ground on another phase constitutes a short circuit.

In the wye system, while the coils have a lower voltage imposed on them, if the system is grounded, a ground (or fault) on any one phase constitutes a short circuit. This may cause the circuit to be deenergized with outages to consumers supplied from the circuit.

If the system is not grounded, a ground (or fault) will establish a "common" grounded point, not symmetrical to the circuit's three phases. The voltages on each phase become distorted. The voltage on the grounded phase may be lower than its nominal values, possibly causing damage from overvoltage to consumers supplied from those two phases.

Considering the supply of power to a motor or engine, DC provides a continuous and even flow, making for a smooth effort. With AC, or push-pull, the flow will build to a maximum and then subside (Fig. C-15). If all the power is supplied in one circuit or one phase, the resulting effort will be very rough. If it can be supplied in three parts or three phases, each one-third in magnitude but applied three times in a given time, the result is less rough. This is similar to hammering home a nail with one big blow, as opposed to using three blows, each one-third in magnitude; the effect is smoother.

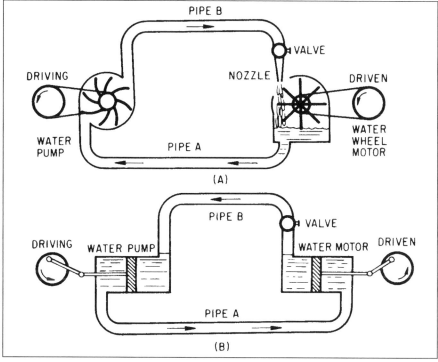

Fig. C-15 The need for using three-phase power: (**a**) smooth flow in one direction analogous to DC power flow; (**b**) reciprocating motion analogous to AC single-phase power flow (More cylinders would be analogous to three-phase power flow.)

Another analogy may be that of a large-bore, two-cylinder engine compared to a small-bore, six-cylinder engine in an automobile. The power of both engines is the same, but the latter drives more smoothly than the former. Hence, three-phase power transmission is usually specified in AC systems.

Comparison of AC Systems

Comparisons of the various AC systems assuming the same (balanced) load, the same voltage between conductors, and the same conductor size are shown in Table C-1 and Table C–2. Using a single-phase, two-wire circuit as a basis for comparison, the relative amount of conductor, power loss, and voltage drop for the different types are shown.

Table C-1 Comparison of AC Systems

Type of System	Conductor Amount	Power Loss	Voltage Drop
Single-phase, 2-wire	1.0	1.0	1.0
Single-phase, 3-wire	1.5	0.25	0.25
Two-phase, 2-wire	1.5	0.50	0.50
Two-phase, 4-wire	2.0	0.50	0.50
Two-phase, 5-wire	2.5	0.50	0.50
Three-phase, 3-wire*	1.5	0.167	0.167
Three-phase, 3-wire**	1.5	0.50	0.50
Three-phase, 4-wire*	2.0	0.167	0.167

*Wye voltage same as single-phase
**Delta voltage same as single-phase
More detail is given in Table C-2, Table of Transmission Efficiencies

Review

• Electrical pressure is measured in volts. Current is given in amperes, resistance in ohms, power in watts, and energy in watt-hours.

• The relationship between these quantities is useful:

$$\text{Ohm's Law: (Current amperes)} = \frac{\text{Pressure (volts)}}{\text{Resistance (ohms)}}$$

$$\text{Power (watts)} = \text{Pressure (volts)} \times \text{Current (amperes)}$$

$$\text{Energy (watt-hours)} = \text{Power (watts)} \times \text{Time (hours)}$$

• Electricity requires a complete circuit to flow. The two kinds of circuits are the series circuit and the multiple or parallel circuit. The series-parallel circuit is a combination of the two.

• In a series circuit, the same current flows in each of the components. In a multiple circuit, the voltage across each of the components is the same.

• Power is the rate of expending energy. Energy is the expenditure of power over a period of time.

Table C-2 Transmission Efficiencies

Type of Circuit		Current in Conductor	Power Loss (I^2R) in Conductor	Voltage Drop (IR)
1. Single-phase, 2-wire		I_1	$I_1^2 \times 2R = 2I_1^2R$	$I_1 \times 2R = 2I_1 R$
2. Single-Phase, 3-wire		$I_2 = \frac{1}{2}I_1$	$\begin{aligned} I_2^2 \times 2R &= (\tfrac{1}{2}I_1)^2 \times 2R \\ &= \tfrac{1}{2}(2I_1^2R) \end{aligned}$	$\begin{aligned} I_2 \times R &= \tfrac{1}{2}I_1 \times R \\ &= \tfrac{1}{4}(2I_1 R) \end{aligned}$
3. Two-Phase, 4-wire		$I_{3a} = \frac{1}{2}I_1$ $I_{3b} = \frac{1}{2}I_1$	$\begin{aligned} I_3^2 \times 2R \times 2 &= 4I_3^2 \times R \\ &= 4(\tfrac{1}{2}I_1)^2 \\ &= \tfrac{1}{2}(2I_1^2R) \end{aligned}$	$\begin{aligned} I_3 \times 2R &= 2I_3 R \\ &= 2(\tfrac{1}{2}I_1)R \\ &= \tfrac{1}{2}(2I_1 R) \end{aligned}$

Table C-2 Transmission Efficiencies (continued)

4. Two-Phase, 3-wire

$$I_4 = \tfrac{1}{2}I_1$$

$$
\begin{aligned}
I_4^2 &= 2R + \left(\sqrt{2}I_4\right)^2 R \\
&= I_4^2(2R + 2R) \\
&= 4I_4^2 R \\
&= 4(\tfrac{1}{2}I_1)^2 R \\
&= \tfrac{1}{2}(2I_1^2 R)
\end{aligned}
$$

$$
\begin{aligned}
I_4R &+ \sqrt{2}I_4R \\
&= I_4R + 1.41\ I_4R \\
&= 2.42\ I_4R \\
&= 2.42(\tfrac{1}{2}I_1)R \\
&= 1.21 \times I_1R \\
&= \text{approx } \tfrac{1}{2}(2I_1)R
\end{aligned}
$$

5. Two-Phase, 5-wire

Since loads are balanced, no current will flow in the fifth or neutral wire. Hence, current, power loss, and voltage drop will be the same as for the Two-Phase, 4-wire system; refer to item 3 above.

6. Three-Phase, 3-wire—Y

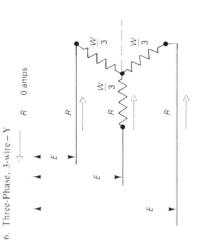

$$I_6 = \frac{I_1}{3}$$

$$
\begin{aligned}
I_6^2 &\times R \times 3 \\
&= \left(\frac{I_1}{3}\right)^2 \times R \times 3 \\
&= \frac{1}{3}I_1^2 R \\
&\text{or } \frac{1}{6}(2I_1^2R)
\end{aligned}
$$

$$
\begin{aligned}
I_6 \times R &= \frac{I_1}{3} = R \\
&= \frac{1}{6}(2I_1R)
\end{aligned}
$$

Table C-2 Transmission Efficiencies (continued)

Type of Circuit	Current in Conductor	Power Loss (I^2R) in Conductor	Voltage Drop (IR)
7. Three-Phase, 3-wire	$I_2 = \dfrac{I_1}{\sqrt{3}}$	$\begin{aligned} I_2^2 &\times R \times 3 \\ &= \left(\dfrac{1}{\sqrt{3}}\right)^2 \times R \times 3 \\ &= I_1^2 R \\ &\text{or } \tfrac{1}{2}(2I_1^2 R) \end{aligned}$	$\begin{aligned} I_2 \times R &= \dfrac{I_1}{\sqrt{3}} R \\ &= \dfrac{1}{2\sqrt{3}}(2I_1 R) \\ &\quad * \end{aligned}$

*For comparison, this value should be multiplied by $\sqrt{3}$ because line to neutral voltage in this case is only $\dfrac{1}{\sqrt{3}}$ of E assumed in case 1. Therefore, comparative voltage drop $= \tfrac{1}{2}(2I_1 R)$

8. Three-Phase, 4-wire — Y

Since loads are balanced, no current will flow in the fourth or neutral wire. Hence, current, power loss, and voltage drop will be the same as for Three-Phase, 3-wire — Y system; refer to item 6.

*Based on fixed, balanced loads at power factor of 1.0 and same wire size.

- Current in a wire produces heat and a loss or drop in voltage.

- Inductance is the obstruction to the flow of current in an AC circuit caused by magnetic lines of force cutting it. These lines of force may be produced by the current flowing in the conductor itself, or by an adjacent conductor. The first is called self-inductance, the latter mutual inductance. They are expressed in ohms.

- Capacitance is the obstruction to the flow of current in an AC circuit caused by electrostatic fields set up by adjacent conductors; it is expressed in ohms.

- Inductance and capacitance in an AC circuit cause the current and voltage to be displaced from each other and not to act together when producing power.

- Impedance is the net result of the action of resistance, inductance, and capacitance in an AC circuit; it is expressed in ohms. Ohm's Law for AC becomes:

$$\text{Current (amperes)} = \frac{\text{Pressure (volts)}}{\text{Impedance (ohms)}}$$

$$\text{Power (watts)} = \text{Pressure (volts)} \times \text{Current (amperes)} \times \text{Power factor (percent)}$$

- Power factor is the ratio of real power to apparent power; it is expressed in percent.

- The effective value of an AC sine-wave voltage or current is 70.7% of the maximum value of the sine wave and is equal to the effective value of DC voltage or current values.

- The basic circuit (either DC or AC) consists of two wires, a sending one and a return.

• In an AC circuit, power may be delivered over two or three circuits or phases whose voltages are displaced electrically by 120° for three-phase circuits.

Study Questions

1. What are the units of electrical pressure, current, and resistance?
2. Express the relationship between the three quantities in a simple electric circuit. What name is given to this relationship?
3. What are the two basic types of electric circuits? How do they differ?
4. What is the difference between power and energy?
5. What two factors must be considered in transmitting power a great distance?
6. How may a voltage be induced in a conductor? On what factors does the magnitude of the induced voltage depend?
7. What is meant by inductance? What is meant by capacitance? What effect do they have on the relationship between voltage and current in a circuit?
8. What is meant by impedance? How is Ohm's Law affected by alternating currents? What is meant by resonance?
9. What is meant by power factor in an AC system? How is it expressed?
10. What is meant by a single-phase circuit? By a polyphase circuit? What are the advantages and disadvantages of each?

Appendix

Electrical Power Glossary

Accuracy. The extent to which a given measurement agrees with a standard defined value. For revenue metering, public service authorities require stringent, very high degrees of accuracy in watt-hour and demand meters. These include pre-installation testing and calibration as well as periodic testing, which may be done on a sampling basis for electric utilities. Remote meter reading devices must also meet both the utility standard and public service authority standards for accuracy.

ACSR. Aluminum cable, steel reinforced.

Adjustment clauses. A clause or series of clauses in a rate schedule (tariff) that permits a utility to charge for changes in fuel costs, temperature deviations from normal, cost of purchased power, ratchet demands, etc.

Allocation. The procedural step in a cost of service study whereby joint costs are allocated among consumer classes based on demand, energy, or some other cost-related feature of service.

Allowable costs. Expenses that are allowed as operating costs chargeable to the consumers; certain other costs are chargeable to the stockholders or other owners.

207

Ammeter. An instrument to measure current flow, usually indicating in amperes. Where current is measured in milliamperes (1/1,000 of an ampere), the instrument may be called a milliammeter.

Ampere (amp). The unit of measurement of electric current. It is proportional to the quantity of electrons flowing through a conductor past a given point in 1 second. It is the unit current produced in a circuit by 1 volt acting across a resistance of 1 ohm.

ASE cable. A variant of SE cable in which a flat, steel strip is inserted between the neutral conductor and the outside braid for greater mechanical strength.

Automatic meter reading. A method of reading a meter (watt-hour, demand, gas, water, or any other type of meter), preparing and conditioning the data, and transmitting the accumulated information from the meter location back to a central data accumulation device. This central collection device in most cases is some form of computer. The communication link may be radio, telephone line, power line carrier, cablevision, or any combination thereof.

Average annual electric bill. Annual electric revenue from a class of service divided by the average number of such consumers for the 12-month period.

Average cost. A method of determining the cost of providing service to the various consumer classes. Average cost-of-service figures may be used in setting rates. Average costs are total costs divided by the number of units produced. This method, while distinguishing costs between different consumer classes, fails to recognize that not all kilowatts and kilowatt-hours are produced at the same cost within one consumer class. Seasonal, time-of-day, and marginal cost-based rates more accurately reflect the true costs of producing each kilowatt or kilowatt-hour.

Avoided costs. The costs an electric utility would otherwise incur to generate power if it did not purchase power from another source.

Basic reference standards. Those standards with which the value of electrical units are maintained in a laboratory, and which serve as the starting point of the chain of sequential measurements carried out in the laboratory.

Bolometer. A device used for measuring heat at a distance; used to patrol electrical lines and locate hot spots before failure occurs.

Bottom-connected meter. A meter having a bottom-connection terminal assembly. Also referred to as an A-base meter.

Capability. In general, the maximum load in amperes, kW, or kVA that a system or a component of a system can carry without exceeding its design limits. They are usually defined as being normal or emergency limits. Normal limits are those that can meet expected conditions in the normal operation of the system without damage to the facility or without incurring a significant loss of expected life of the item. Emergency capability is some level above the normal limit that takes into account specific conditions of ambient temperature, preceding loading, and a calculated loss of life of the equipment.

Capacitance. That property of an electric circuit that allows storage of energy and exists whenever two conductors are in close proximity but separated by an insulation or dielectric material. When direct voltage is impressed on the conductors, a current flows momentarily while energy is being stored in the dielectric material. It stops when electrical equilibrium is reached. With an alternating voltage between the conductors, the capacitive energy is transferred to and from the dielectric material, resulting in an alternating current flow in the circuit.

Central station. Control equipment, typically a computer system, which can communicate with metering and load-control devices. The equipment may also interpret and process data, accept data from other sources, and prepare reports or consumer bills.

Class designation. The maximum of the watt-hour meter load range in amperes.

Connection charge. An amount to be paid by the consumer in a lump sum, or in installments, for connecting a consumer's facilities to the electric system.

Constant kilowatt hour. Pertaining to a meter (register constant, dial constant): the multiplier applied to the register reading to obtain kilowatt-hours.

Corona discharge. A luminous effect from the discharge of electrical energy on the surface of a conductor. Corona discharge occurs when the conductor carries voltage exceeding a critical value.

Creep. For mechanical meters, a continuous motion of the rotor of the meter with normal operating voltage applied and the load terminals open-circuited. For electronic meters, a continuous accumulation of data in a consumption register when no power is being consumed.

Demand. The rate at which electric energy is delivered to or by a system, part of a system, or a piece of equipment. It is expressed in kilowatts, kilovolt-amperes, or other suitable unit at a given instant or averaged over any designated period of time. The primary source of demand is the power-consuming equipment and devices of consumers.

Demand charge. That portion of the charge for electric service based on the peak load furnished within a time period according to the established tariff.

Demand constant. Pertaining to pulse receivers, the value of the measured quantity for each received pulse, divided by the demand interval, expressed in kilowatts per pulse, kilovars per pulse, or other suitable units. The demand interval must be expressed in parts of an hour such as 1/4 for a 15-minute interval or 1-1/2 for a 5-minute interval.

Demand interval. The period of time during which the electric energy flow is averaged in determining demand, such as 60 minutes, 30 minutes, 15 minutes, or instantaneously.

Demand meter. A metering device that indicates or records the demand and/or the maximum demand, or both. Since demand involves both an electrical factor and a time factor, mechanisms responsive to each of these factors are required, as well as an indicating or recording mechanism. These mechanisms may be either separate from or structurally combined with one another. An alternative mode would be to have a computer interpret and calculate the desired demand.

Demand register. A mechanism (for use with an integrating electricity meter) that indicates maximum demand and also registers energy (or other integrated quantity).

Demand side management (DSM). The broad term for electric load management applying to utility actions, programs, and designs for the purpose of lowering system peaks and reducing energy consumption by the consumer as well as the utility system. There are direct actions and controls by the utility that result in immediate results to lower peaks and reduce energy requirements. There are also indirect (passive) programs requiring the cooperation and participation by the consumer to achieve similar results.

Detent. A device installed in a meter to prevent reverse rotation.

Dial-out capability. The ability of a meter to initiate communications with a central station, usually using telephone lines.

Disk constant. See watt-hour constant.

Disk position indicator. Also known as a "caterpillar," it is an indicator on the display of a solid state register that simulates rotation of a disk at a rate proportional to power.

Diversity. That characteristic of a variety of electric loads whereby individual maximum demands usually occur at different times. This permits design of capability to meet a demand reduced from the sum of all the individual peak demands.

Dump energy. Energy generated that cannot be stored or conserved when such energy is beyond the immediate use in a system. This energy is usually bid out to neighboring systems at a price less than the cost to produce it.

Electromagnet. A magnet in which the magnetic field is produced by an electric current. A common form of electromagnet is a coil of wire wound on a laminated iron core, such as the voltage coil of a watt-hour meter stator.

Embedded coil. A coil in close proximity to, and nested within, a current loop of a meter to measure the strength of a magnetic field and develop a voltage proportional to the flow of current.

Encoder. A device that converts a meter reading into a form suitable for communicating to a remote central location or to a portable recording device or remote dial.

Energy charge. That portion of the charge for electric service based on the energy (kWh) consumed by the consumer.

Energy conservation. The strategy of a utility leading to programs for the reduction of electric energy consumption by the consumer. This includes rebates and/or low-cost financing assistance for the consumer to install or replace appliances, lighting, and motors with more efficient ones. It also includes using higher grades of insulation to reduce heat loss in cool weather and to reduce heat absorption in warm weather. Utilities will often furnish information and technical assistance at no cost to the consumer, but it is the consumer who will make the decision to increase his capital cost to reduce his energy costs. In addition to reducing energy consumption, conservation also lowers the peak loads.

Functional accounts. Groupings of plant and expense accounts according to the specific function in the electric system, sometimes referred to as the Uniform System of Accounts. For instance, there would be an account number such as 13804 or some other specific number for the capital value of all the meters a utility owns.

Gear ratio. The number of revolutions of the rotating element of a meter for one revolution of the first dial pointer.

Grounding conductor. A conductor used to connect any equipment device or wiring system with a grounding electrode or ground system.

Incentive rate. Some form of reduced rate generally designed to provide an incentive for targeted consumers to remain in the service territory or to lure other businesses to the territory, usually offered for a fixed period of time. Time-of-day rates are in some sense an incentive rate to the consumer to time his load use so as to minimize his energy or demand use during peak load periods.

Inductance. That property of an electric circuit that opposes any change of current direction through the circuit. In a direct current circuit, where current does not change in direction, there is no inductive effect. In alternating current (AC) circuits, the current is constantly changing direction, so the inductive effect is appreciable. Changing current produces changing flux, which, in turn, produces induced voltage. The induced voltage opposes the change in applied voltage, hence the opposition to the change in current. Since the current changes more rapidly with increasing frequency, the inductive effect also changes with frequency.

Instrument transformer. A transformer that reproduces in its secondary circuit, in a definite and known proportion, the voltage or current of its primary circuit, with the phase relationship substantially preserved.

IPP. Independent power producer.

ISO. Independent system operator.

K_e. *See* **KYZ output constant.**

K_h. *See* **watt-hour constant.**

K_m. *See* **mass memory constant.**

kVA. One thousand volt-amperes

kW. One thousand watts.

kWh. One thousand watt-hours.

KYZ output constant (K_e). The pulse constant for the KYZ outputs of a solid state meter, programmable in unit-hours per pulse.

Lagging current. An alternating current that, in each half cycle, reaches its maximum value a fraction of a cycle later than the maximum value of the voltage that produces it.

Leading current. An alternating current that, in each half cycle, reaches its maximum value a fraction of a cycle sooner than the maximum value of the voltage that produces it.

Load compensation. That portion of the design of a watt-hour meter that provides good performance and accuracy over a wide range of loads. In modern, self-contained meters, this load range extends from load currents under 10% of the rated meter test amperes to 667% of the test amperes for class 200 meters.

Load control. A direct action by the utility to shift load off-peak, either by a programmed computer command or by manual implementation of remote control. Voltage reduction and cycling of motor and appliance loads and water/house heating are examples of load control.

Load curve. A curve on a chart showing energy, power, or amperes plotted against time of occurrence to show the varying magnitude of a load during the period covered.

Load factor. The ratio of the average load in kilowatts (kW) supplied during a designated period to the peak or maximum load in kW occurring in that period. Load factor, in percent, also may be derived by multiplying the kilowatt-hours in the period by 100 and dividing by the product of the maximum demand in kW and the number of hours in the period.

Load forecast. A predicted demand or energy amount expected during a period of time or at a specific instant in time. Load forecasts may be short-term for operating purposes, long-term for system planning purposes, or any range in between. Forecasts may be of total system or regional load, or of areas of the system such as served by a substation or distribution circuit. They might also be of consumer loads (especially in the case of transmission consumers), or appliance/device loads, including street lighting.

Load research. Generally refers to a utility activity designed and carried out to determine consumer load characteristics. The results of load research activities are not only used in rate analysis and development but also to determine the electric design parameters, both for the system and the components of the system.

Load shape control. *See* **load shifting**.

Load shifting. Reducing the system peaks by curtailing load directly or by a rate structure that motivates the consumer to move some of his energy requirements from peak load hours to off-peak usage. Filling the dips in the load curve improves the system load factor and makes more use of lower cost energy produced by base load units or off-peak purchased power.

Load survey. The measurement of the electrical characteristics of a consumer or segment of the electric system. This is usually done by portable special load monitoring instruments installed either on the consumer premises or on some part of the electric supply system to measure demand, energy usage, amperes, voltages, and/or power factor. An alternative to special field instrumentation would be to transmit consumer data back to a central location and develop the required information through computer programs.

Loss compensation. A means for correcting the reading of a meter when the metering point and point of service are physically separated, resulting in measurable losses. These losses include I^2R losses in conductors and transformers, and iron core losses. These losses may be added to, or subtracted from, the meter registration.

Loss factor. The ratio of the average loss in kilowatts during a designated period to the peak, or maximum loss of kilowatts occurring in that period.

Losses. The general term applied to energy (kilowatt-hours) and power (kilowatts) lost in the operation of an electric system. Losses occur principally as energy transformations from kilowatt-hours to waste heat in electrical conductors and equipment or devices.

Mass memory constant (K_m). The value, in unit quantities, of one increment (pulse period) of stored serial data. For example, K_m = 2,500 W-hr/pulse.

Maximum demand. The demand usually determined by an integrating demand meter or by the integration of a load curve. It is the summation of the continuously varying instantaneous demands during a specified time interval. With the computer technology available, the integration of demand can also be accomplished through automatic meter readings transmitted from the consumer meter back to a central location. This makes it possible not to have a demand meter or register at the location of usage.

Memory. Electronic devices that store instructions and data. Volatile memories can be written to, and read from, repeatedly. Random-access memories (RAM) require uninterrupted power to retain their contents. Read only memories (ROMs) are programmed once, may only be read (but can be read repeatedly), and do not require constant power to retain their contents. ROMs are typically used to store firmware in dedicated systems.

Ohm. The practical unit of electrical resistance. It is the resistance that allows 1 ampere to flow when the impressed electrical pressure is 1 volt.

On-peak demand register. A register that will record the total energy used and, in addition, will register maximum demand during on-peak periods. The control of the demand recording is with a solenoid-operated demand gear train that may be actuated from a local device or from a device located remotely.

Percent registration. Percent registration of a meter is the ratio of the actual registration of the meter to the true value of the quantity measured in a given time, expressed as a percentage. Percent registration is also sometimes referred to as the accuracy of the meter.

Phantom load. A device that supplies the various load currents for meter testing, used in a portable form for field testing. The power source is usually the service voltage, which is transformed to a low value. The load currents are obtained by suitable resistors switched in series with the isolated low-voltage secondary and output terminals. The same principle is used in most meter test boards.

Phase angle. The phase angle or phase difference between a sinusoidal voltage and a sinusoidal current is defined as the number of electrical degrees between the beginning of the cycle of voltage and the beginning of the cycle of current.

Photoelectric meter. Also known as a counter, this device is used in the shop testing of meters to compare the revolutions of a watt-hour meter standard with a meter under test. The device receives pulses from a photoelectric pickup, which is actuated by the anticreep holes in the meter disk (or the black spots on the disk). These pulses are used to control the standard meter revolutions on an accuracy indicator by means of various relay and electronic circuits.

Power factor (PF). The ratio of real power (kW) to apparent power (kVA) at any given point and time in an electrical circuit. Generally it is expressed as a percentage ratio. Resistance-type loads, such as incandescent lamps or electric resistance heating, are characterized as 100% PF loads. Electric motor loads in refrigerators and air conditioners are around 80% PF or lower.

Primary/transmission metering. When consumer loads reach a magnitude such that they will benefit from higher service voltage rates, the revenue will be metered by special metering installations called primary metering. This means the consumer is served at a distribution voltage of 4,000 volts (V) or higher (usually 13,000 V) or by a transmission line, usually operating at 69,000 V or higher. In these cases, the consumer furnishes his own step-down transformers and high-voltage service equipment. The utilities use potential transformers (PTs) and current transformers (CTs) to reduce both potential and current to levels for which the meter is designed.

Pulse device. For electric metering, it is the functional unit for initiating, transmitting, retransmitting, or receiving electric pulses, representing finite quantities, such as energy. Normally pulses are transmitted from some form of electric meter to a receiver unit.

Pulse initiator. Any device, mechanical or electrical, used with a meter to initiate pulses, the number of which is proportional to the quantity being measured. It may include an external amplifier or auxiliary relay or both.

Q-hour meter. An electricity meter that measures the quantity being obtained by lagging the applied voltage to a watt-hour meter by 60°, or for electronic meters by delaying the digitized voltage samples by a time equivalent to 60° (electrical).

Rate base. The value established by a regulatory authority, upon which a utility is permitted to earn a specified rate of return. Generally this represents the amount of property used and useful in servicing of consumers.

Reactance. The measurement of opposition to the current flow in an electric circuit caused by the circuit properties of inductance and capacitance, normally measured in ohms.

Reactive power. The portion of apparent power that does no work. It is commercially measured in kilovars (volt-amperes reactive). Reactive power must be supplied to most types of magnetic equipment, such as motors. It is supplied by generators or by electrostatic equipment known as capacitors.

Real power. The energy or work-producing part of apparent power; the rate of supply of energy, measured commercially in kilowatts. The product of real power and length of time is energy, measured by watt-hour meters and expressed in kilowatt hours (kWh).

Register constant. The number by which the register reading is multiplied to obtain kilowatt-hours. The register constant on a particular meter is directly proportional to the register ratio, so any change in ratio will change the register content.

Register freeze. The function of a meter or register to make a copy of its data, and perhaps reset its demand, at a pre-programmed time after a certain event (such as demand reset) or upon receipt of an external signal. It is also known as self-read, auto-read, or data copy.

Register ratio. The number of revolutions of the gear meshing with the worm or pinion on the rotating element for one revolution of the first dial pointer.

Registration. The registration of the meter is equal to the product of the register reading and the register constant. The registration during a given period of time is equal to the product of the register constant and the difference between the register readings at the beginning and the end of the period.

SE cable. Service entrance cable usually consists of at least two conductors with appropriate insulation, laid together and covered with a jacket material about which is wrapped a stranded, bare neutral conductor. An outer covering over the neutral conductor is a flame-retarding and waterproof braid. (*See also* **ASE cable**.)

Service entrance conductors. For an overhead service, that portion of the service conductors that connects the service drop to the service equipment. The service entrance for an underground service is that portion of the service conductors between a terminal box (external or internal) and the service equipment. In the absence of a terminal box, the service entrance runs from the point of entrance into the building to the service equipment junction.

Skin effect. The tendency of current flowing within a conductor to flow more easily (and in greater part) near the surface of the conductor in an AC circuit.

Socket. Also known as a trough, it is the mounting device consisting of jaws, connectors, and enclosure for socket-type meters. A mounting device may be either a single socket or a trough. The socket may have a cast or formed enclosure, the trough an assembled enclosure which may be extended to accommodate more than one mounting unit, as may be encountered in multiple occupancy buildings.

Stator. The unit that provides the driving torque in a watt-hour meter. It contains a voltage coil, one or more current coils, and the necessary steel to establish the required magnetic paths.

Submetering. The metering of individual loads within a building or subdivisions of property (such as a trailer park) for billing purposes by the owner. For that application, usually there is one master meter by the utility for billing the owner and the sub-meters are used by the owner to charge tenants for energy used.

Temperature compensation. In reference to a watt-hour meter, refers to the factors included in the design and construction of a meter that make it perform with good accuracy in a wide range of temperatures. In modern electric meters this may range from –20° F to 140° F. In a gas meter installation, this temperature compensation may also mean the factor applied to compensate for the difference in temperature between inside metering and outside metering.

Three-rate watt-hour meter. A watt-hour meter with three sets of registers. It is constructed so that the off-peak energy will be recorded on one set of dials, and the on-peak energy for two different on-peak periods will be recorded on the other two sets of dials. The control of the recording system is by an internal switch or remote control system. In the event of a power failure, carryover for an internal time switch can be accomplished by battery or spring storage.

Time division multiplication. An electronic measuring technique that produces an output signal proportional to two inputs, for example, voltage and current. The width or duration of the output signal is proportional to one of the input quantities; the height is proportional to the other. The area of the signal is then proportional to the product of the two inputs.

Time-of-day rates. A rate strategy enacted by the utility to influence the consumer to change his load pattern and shift loads to off-peak hours by charging a higher rate for usage during peak hours and lower rates for off-peak hours.

Time-of-use metering. A metering method that records consumption and demand during selected periods of time. This allows consumption and demand (in some cases for commercial loads or very large houses) to be billed at different rates established in the utility tariff.

Transducer. A solid-state electronic device to receive energy from one system and supply energy, of either the same or different kind, to another system. This occurs in such a manner that the desired characteristics of the energy input appear at the output. For instance, it may take an AC voltage frequency or magnitude from a suitable interface and transmit it to a reading device or control relay.

Trough. *See* **socket.**

Two-rate watt-hour meter. A watt-hour meter with two registers or sets of dials, constructed so that the off-peak energy will be recorded on one set of dials and the on-peak energy on the other set. The control of the recording system is by an internal switch or external remote control signal. In the event of a power failure, carryover for an internal time switch can be accomplished by battery or spring storage.

V. *See* **volt.**

VA. *See* **volt-ampere.**

Var-hour meter. An electricity meter that measures and registers the integral, with respect to time, of the reactive power of the circuit in which it is connected. The unit in which this integral is measured is usually the kilovar-hour. This is most often used to obtain power factors.

Volt. The unit of electromotive force or electric pressure. It is the electromotive force that, if steadily applied to a circuit having a resistance of 1 ohm, will produce a current of 1 ampere; abbreviated V.

Voltage control. A utility's actions to reduce peak load, either on a planned basis or in an emergency situation in which generation or tie capacity is not available to meet the demand. A signal from a system operator to a remote device at substations automatically reduces the outgoing primary voltage and results in a temporary immediate load reduction.

Voltage of a circuit. The electric pressure of a circuit or system measured in volts. It is generally a nominal rating based on the maximum normal effective difference of potential between any two conductors of a circuit. In a typical house service and meter, this nominal voltage of 120 volts (V) can fluctuate between 117 V and 124 V due to consumer load variations.

Volt-ampere. The basic unit of apparent power. The volt-amperes (VA) of an electric circuit is the mathematical product of the volts and amperes of the circuit. The practical unit of apparent power is the kilovolt-ampere (kVA), which is 1,000 volt-amperes. An average residential service may have an apparent power of 5–10 kVA compared to a primary distribution circuit capability of 8,000 kVA and a transmission circuit capability of 100,000–700,000 kVA.

Watt. The electrical unit of real power or rate of doing work. The rate of energy transfer flowing due to an electrical pressure of 1 volt at unity power factor. One watt is equivalent to approximately 1/746 horsepower, or 1 joule per second. An average size incandescent light bulb uses 75 watts.

Watt-hour. The total amount of energy used in 1 hour by a device that requires 1 watt of power for continuous operation. Electric energy is commonly sold by the kilowatt-hour (kWh; 1,000 watt-hours).

Watt-hour constant (K_h). For an electromechanical meter, the number of watt-hours represented by one revolution of the disk. It is determined by the design of the meter and not normally changed; also called the disk constant. For a solid state meter (K_h or K_t): the number of watt-hours represented by one increment (pulse period) of serial data. Example: K_h or $K_t = 1.8$ watt-hours/pulse.

Watt-hour meter. An electricity meter that measures and registers the integral, with respect to time, of the active power (often referred to as real power) of the circuit in which it is connected. This power integral is the energy delivered to the circuit over which the integration extends, and the unit measured is usually the kilowatt-hour. This is the most frequently used revenue meter device on an electric utility system found in residences, commercial establishments, and industrial plants.

Wheeling charge. An amount paid by a consumer or utility for transporting power over electric lines owned by others. This can be in the form of an energy charge, demand charge, capacity charge, or a combination of these.

Electric Utility Ownership and Operation

Electric utilities can be classified by size, type of operation, ownership, and territory served. While there are no typical or usual organizations in the climate of competition and corporate "re-engineering" evident in the past few years, there are some generalizations that can be made. It should also be remembered that an electric utility may also be in the natural gas business or other commercial endeavors.

Utility Organization

There are broad general function splits between support services and actual operations in the medium-to-large size utilities that may not be found in smaller utilities (see Table E–1). If a utility is large enough, divisions of functions and areas may be established that then report back to a central management group or authority.

Table E–1 Utility Support Services and Operations

Support Services	Operations
legal	generation
finance	system operation
accounting	line construction
public relations	line maintenance
human resources	distribution operation
purchasing	major construction
computer services	emergency (storm) restoration
transportation	
engineering and survey	
load research and rates	

Services Found in Either Category

system planning
meters
meter reading
stores
distribution engineering
mapping
general shops
consumer facility engineering

Utility Classification—Ownership

1. Investor-owned or private utilities can be found in all sizes and in all areas of the United States. These utilities are owned by stockholders who expect a return on their investment. Such utility corporations are not guaranteed a profit but do have a maximum return on their investment set by a public authority, who determines the rate tariffs charged to consumers. These utilities pay both income taxes and property taxes. The largest investor-owned utility in Texas is Texas Utilities Company, Dallas, with a summer peak load of 18,000 MW (1990) serving a population of 5,000,000.

2. A second classification of ownership is the Rural Electric Membership Corporation (REMC) set up by the Federal Rural Electrification Administration (REA). These were originally set up to bring electric energy to the more rural areas of the country but now also serve fair-sized towns. They frequently will have the designation "cooperative"

or "association" associated with their geographic name. One of the largest in Texas is the Pedernales Electric Cooperative, Johnson City, with a summer system peak load of 537 MW (1990) and 81,000 consumers.

3.   A third class of ownership consists of the municipal utilities, which are usually owned by the political area served. As of this writing, they do not pay taxes as investor-owned utilities do and frequently enjoy lower cost of financing because of their public ownership and ability to raise money through taxes on their constituents. Most municipals are small in size, although some urban "munis" are large in scale, such as the Austin Electric Utility Department with a summer peak load of 1,483 MW (1990) and 265,000 consumers.

4.   The fourth class of ownership consists of federal, state, and district systems. These are similar to "munis" in that they pay no taxes and have very low rates of financing. They tend to cover large geographic areas and are generally wholesalers to other electric utilities. Well-known examples are the Tennessee Valley Authority (TVA), Bonneville Power Authority (BPA), and the New York Power Authority (NYPA). A state-owned Texas example is the Lower Colorado River Authority, operating out of Austin with a net peak load of 2,257 MW in 1997.

5.   Overlaying the individual utilities are operating entities known as "power pools" that coordinate interchange of power between individual companies and between other power pools. Regional groups of power pools are often organized as coordinating councils, such as the Middle Atlantic Coordinating Council (MACC). These regional councils, as well as the individual power pools, establish area transmission and generation plans, operating standards, and pricing of energy. There is also environmental involvement through committees made up of representatives of the individual companies.

Appendix F

The Electric Utility System

The typical electric utility system is comprised of generation, transmission, substation, and distribution components. They are operated together to supply electrical energy to consumers at any given time and in any given amount of energy required (see Fig. F–1).

Generation

Generation consists of the electrical energy producing or manufacturing facilities owned by the utility company and those facilities owned by others that are connected to the electric utility system. The latter may be in the form of refuse recovery plants, cogeneration, or separately owned generating plants. A utility may also own all or part of a generating plant located remotely from its own territory.

Generating facilities are usually designated as either base load generation or peaking generation. Base load generation units are generally the most economical energy-producing units that are run to supply a major portion of the system load for most of the time. Peaking generation units are ones that are run to make up the difference between base load and the maximum loads either during daily load cycles or seasonally.

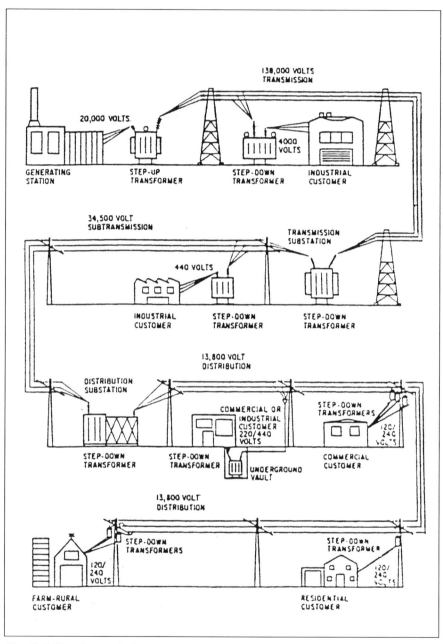

Fig. F–1 Electric supply from generator to customer

Transmission

The transmission part of the electric system moves the power from generating plants and interconnections with other utilities to load centers throughout the utility territory. Transmission lines operate at voltages up to 750,000 volts (V) AC and may be overhead on structures or underground, and in some special cases, under water.

Substations

At load centers in the utility territory, the transmission voltage is reduced (stepped down) to a primary distribution voltage or in some cases to a subtransmission voltage. This allows power to be brought to consumers in the local area by the primary distribution system.

Primary distribution

From the substation, power is distributed to the local area over primary voltage distribution lines operating at voltages from 4,000 V to as many as 45,000 V. The most frequently used primary voltage is 13,000 V. As in the case of transmission, distribution lines may be overhead or underground, but it is much more frequently used underground than in transmission systems. This is particularly true in urban areas and high load density locations, such as shopping centers or industrial parks.

Distribution transformers

A distribution transformer is used to reduce the primary voltage to a utilization voltage that consumers can use within their premises. Residential consumers are usually served at 240/120 V. Appliances such as large air conditioners, electric ranges, and pumps utilize 240 V, while lighting, TVs, and computers use 120 V. Commercial and industrial consumers use higher utilization voltages (up to 480 V).

Appendix *G*

Meter Reading

Most utility revenue meter reading is done manually by going to a consumer location, reading the meter, recording it in some fashion, and returning the data to a central location. Some (a relatively small portion) meters are read by automatic devices that transmit data back to a central control, recording, and processing unit.

Meter Routes

A block of meters is set up for a meter reader to be able to cover in a normal working day in a relatively small geographic area. These vary greatly with the characteristics of the area; in rural, mountainous, desert, or farm areas, the meters will have considerable distance between them compared to a suburban housing development of single-family homes, condos, or townhouses. Urban apartments, shopping centers, and other high-density locations also affect the meter route.

Billing Cycles

A common practice is to read meters every other month and estimate the bill in between using the consumer use characteristic. Commercial and industrial meters are usually read every month because of their high energy use and demands, and because they usually represent a small percentage of the total meters on a system. Utilities have also made available to residential consumers a budget payment plan, so they pay an estimated average fixed amount each month until the end of the budget period. At the end of the period the actual cost is reconciled with the total of the budget payments.

Estimated Reading

An estimated reading can be made by using the historical meter reading information, generally contained in computer files. This can be done for the different seasons and adjusted for any trend indicated by load additions or reductions from improving consumer energy efficiencies.

Hard-to-Read Meters

Inside meter locations present the most common impediment to physically reading a meter, because it depends on someone being there to let the meter reader in. This can be costly to make an actual read if several visits are needed or special appointments have to be set up to get an actual reading. There may be, in some cases, a public service requirement to have an actual read within a certain time period. Some utilities utilize a remote meter dial mounted outside the house, which is connected to the inside meter register. Other "hard-to-reads" are caused by physical obstructions and animals. The possible danger of physical harm to a meter reader inherent in some urban situations also comes under the hard-to-read category.

Cut-Ons and Cut-Offs

When ownership or tenant of a premise changes, the utility may remove the meter or discontinue the service temporarily to show the change in consumer accounts. Cut-ons are more usual for new construction, while a cut-off can also be caused by nonpayment of the bill or theft of service. The latter requires a physical action in the field to actually change the service status by removing the meter or disconnecting the service wires at the pole and/or to read and inspect the meter.

Theft of Service

A residential meter can sometimes be removed and reinserted so that the register runs in reverse, or the meter removed and a jumper installed for a period of time to reduce the actual registration of energy used. To prevent this, most utilities use a meter seal having a unique registered number that is broken if the meter is removed for any reason. If an unauthorized meter removal is made, this would be detected by the meter reader or some other operating person. Programmed computer checks can also be made to show less-than-normal estimated usage and trigger an investigation. In remote automatic meter reading systems, the theft of service attempt would be detected at the time it occurs.

Appendix *H*

Porcelain vs. Polymer Insulation

For many years, porcelain for insulation purposes on lines and equipment has exercised a virtual monopoly. It was perhaps inevitable that plastics1 successful as insulation for conductors since the early 1950s, should supplant porcelain as insulation for other applications in the electric power field.

The positive properties of porcelain are chiefly its high insulation value and its great strength under compression. Its negative features are its weight (low strength to weight ratio) and its tendency to fragmentation under stress. Much of the strength of a porcelain insulator is consumed in supporting its own weight. (Figs. H-1a and b)

In contrast, the so-called polymer not only has equally high insulation value, but acceptable strength under both compression and tension. It also has better water and sleet shedding properties, hence handles contamination more effectively, and is less prone to damage or destruction from vandalism. It is very much lighter in weight than porcelain (better strength to weight ratio), therefore more easily handled. (Figs H-2; Table H-1)

Economically, costs of porcelain and polymer materials are very. competitive, but the handling factors very much favor the polymer.

Polymer insulation is generally associated with a mechanically

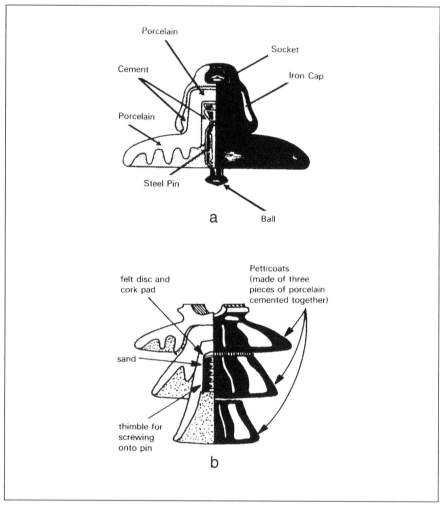

Fig H-1a Ball and socket type suspension insulator. **b** Pin type insulator

stronger insulation, such as high strength fiberglass. The fiberglass insulation serves as an internal structure around which the polymer insulation is attached, usually in the form (and function) of petticoats (sometimes also referred to as bands, watershedders; but for comparison purposes, however, here only the term petticoat will be used). The insulation value of the polymer petticoats is equal to or greater than that of the fiberglass to which it is attached.

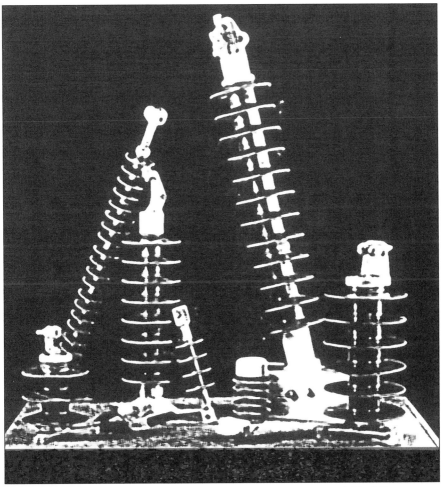

Fig H-2 A variety of typical Polymer insulator shapes (*courtesy Hubbell Power Systems*)

The internal fiberglass structure may take the form of a rod (or shaft), a tube, cylinder, or other shape. It has a high comparable compression strength as a solid and its tensile strength, equally high, is further improved by stranding and aligning around a fiber center. The polymer petticoats are installed around the fiberglass insulation and sealed to prevent moisture or contamination from entering between the petticoats and fiberglass Figure H-3. The metal fittings at either end are crimped directly to the fiberglass, developing a high percentage of the inherent strength of the fiberglass. It should be noted that fiberglass with an elastometric (plastic) covering has been used for insulation purposes since the early 1920s.

Table H-1 Polymer Insulation Weight Advantage

Product	Type	Voltage (kV)	Porcelain Weight (lbs)	Polymer Weight (lbs)	Percent Weight Reduction
Insulator	Distribution	15	9.5	2.4	74.7
Arrester	Distribution	15	6.0	3.8	36.7
Post Insulator	Transmission	69	82.5	27.2	67.0
Suspension	Transmission	138	119.0	8.0	93.2
Intermediate Arrester	Substation	69	124.0	28.0	77.4
Station Arrester	Substation	138	280.0	98.9	64.7

Source: Hubbell Power Systems

The polymer petticoats serve the same function as the petticoats associated with porcelain insulators, that of providing a greater path for electric leakage between the energized conductors (terminals, buses, etc.) and the supporting structures. In inclement weather, this involves the shedding of rain water or sleet as readily as possible to maintain as much as possible the electric resistance between the energized element and the supporting structure, so that the leakage of electrical current between these two points be kept as low as possible to prevent flashover and possible damage or destruction of the insulator. Tables H-2a & b.

Table H-2a Polymer Improvement Over Porcelain

Voltage kV	Watts Loss Reduction* (watts per insulator string) Relative Humidity				
	30%	50%	70%	90%	100%
69	0.6	0.8	0.9	1.0	4.0
138	1.0	2.4	4.5	7.2	8.0
230	1.0	2.5	5.7	14.0	29.0
345	2.5	4.2	8.5	15.0	30.0
500	2.8	7.8	11.5	33.0	56.0

*Power loss measurements under dynamic humidity conditions on I-strings
Source: Hubbell Power Systems

Table H-2b Polymer Distribution Arrester Leakage Distance Advantage

MCOV	Standard Porcelain Leakage Distance (in)	Standard Porcelain Height (inches)	Special Porcelain Leakage Distance (in)	Special Porcelain Height (inches)	Standard Polymer Leakage Distance (in)	Standard Polymer Height (inches)
8.4	9.0	9.4	18.3	15.9	15.4	5.5
15.3	18.3	15.9	22.0	20.0	26.0	8.5
22.0	22.0	20.0	29.0	28.9	52.0	17.2

Source: Hubbell Power Systems

When the insulator becomes wet, and especially in a contaminated environment, leakage currents begin to flow on the surface; if the current becomes high enough, an external flashover takes place. The rate at which the insulation dries is critical. The relationship between the outer petticoat diameter and the core is known as the form factor. The leakage current generates heat (12R) on the surface of the insulator (eddy currents). In addition to the effects of the leakage current, the rate at which the petticoat insulation will dry depends on a number of factors. Starting with its contamination before becoming wet, the temperature and humidity of the atmosphere and wind velocity following the cessation of the inclement weather. In areas where extreme contamination may occur (such as some industrial areas or proximity to ocean salt spray), the polymer petticoats may be alternated in different sizes, Figure H-4, to obtain greater distance between the outer edges of the petticoats across which flashover might occur. When dry, the leakage current (approximately) ceases and the line voltage is supported across dry petticoats, preventing flashover of the insulator. It is obviously

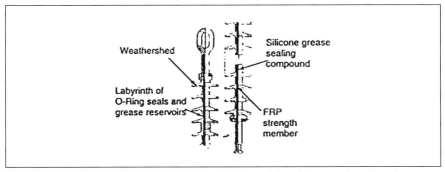

Fig H-3 Polymer insulator—showing fiberglass rod insulation and sealing (*courtesy Hubbell Power Systems*)

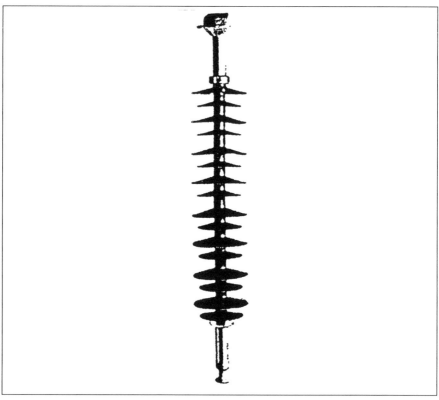

Fig H-4 Polymer insulator arrangement areas of high contamination where flashover between petticoat edges is possible (*courtesy Hubbell Power Systems*)

impractical to design and manufacture comparable porcelain insulators as thin as polymers and having the same form factor. (Table H-3)

Table H-3 Comparison of Contamination Performance of Polymer verses Porcelain Housed Intermediate Class Arresters

MCOV (kV)	Housing Material	Housing Leakage Distance (in)	Max. Current (mA crest)	Max. Disc. Temp. (°C)	5	10	15	20
57	Polymer	81	<1	<38	35	42	44	44
66	Porcelain	54	68	>163	–	–	–	–
84	Polymer	109	<1	<38	50	52	60	60
98	Porcelain	122.4	18	<82	143	160	175	185

Source: Hubbell Power Systems

In addition, in porcelain insulators, the active insulating segment is usually small and, when subjected to lightning or surge voltage stresses, may be punctured In subsequent similar circumstances, it may breakdown completely, not only causing flashover between the energized element and the supporting structure, but may explode causing porcelain fragmentation in the process; the one-piece fiber-glass insulator will not experience puncture.

The polymer suitable for high voltage application consists of these materials:

1.   Ethylene Propylene Monomer (EPM)
2.   Ethylene Propylene Diene Monomer (EPDM)
3.   Silicone Rubber (SR)

Both EPM and EPDM, jointly referred to as EP, are known for their inherent resistance to tracking and corrosion, and for their physical properties. SR offers good contamination performance and resistance to Ultra Violet (UV) sun rays. The result of combining these is a product that achieves the water repellent feature (hydrophobic) of silicone and the electromrchanical advantages of EP rubber.

Different polymer materials may be combined to produce a polymer with special properties; for example, a silicone EPDM is highly resistant to industrial type pollution and ocean salt.

The advantageous strength to weight ratio of polymer as compared to porcelain makes possible lighter structures and overall costs as well as permitting more compact designs, resulting in narrower right-of-way requirements and smaller station layouts. The reduction in handling, shipping, packaging, storage, preparation and assembly, all with less breakage, are obvious—these, in addition to the superior electromechanical performance,

Fiberglass insulation with its polymer petticoats is supplanting porcelain in bushings associated with transformers, voltage regulators, capacitors, switchgear, circuit breakers, bus supports, instrument transformers, lightning or surge arresters, and other applications. The metallic rod or conductor inside the bushing body may be inserted in a fiberglass tube and sealed to prevent moisture or contamination entering between the conductor and the fiberglass tube around which the polymer petticoats are attached. More often, the fiberglass insulation is molded around the conductor, and the polymer petticoats attached in a similar fashion as in insulators. (Fig H-5)

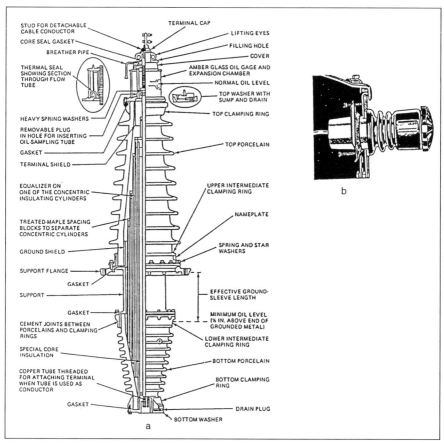

STUD FOR DETACHABLE CABLE CONDUCTOR
CORE SEAL GASKET
BREATHER PIPE
THERMAL SEAL SHOWING SECTION THROUGH FLOW TUBE
HEAVY SPRING WASHERS
REMOVABLE PLUG IN HOLE FOR INSERTING OIL-SAMPLING TUBE
GASKET
TERMINAL SHIELD
EQUALIZER ON ONE OF THE CONCENTRIC INSULATING CYLINDERS
TREATED-MAPLE SPACING BLOCKS TO SEPARATE CONCENTRIC CYLINDERS
GROUND SHIELD
SUPPORT FLANGE
GASKET
SUPPORT
GASKET
CEMENT JOINTS BETWEEN PORCELAINS AND CLAMPING RINGS
SPECIAL CORE INSULATION
COPPER TUBE THREADED FOR ATTACHING TERMINAL WHEN TUBE IS USED AS CONDUCTOR
GASKET

TERMINAL CAP
LIFTING EYES
FILLING HOLE
COVER
AMBER GLASS OIL GAGE AND EXPANSION CHAMBER
NORMAL OIL LEVEL
TOP WASHER WITH SUMP AND DRAIN
TOP CLAMPING RING
TOP PORCELAIN
UPPER INTERMEDIATE CLAMPING RING
NAMEPLATE
SPRING AND STAR WASHERS
EFFECTIVE GROUND-SLEEVE LENGTH
MINIMUM OIL LEVEL (⅛ IN. ABOVE END OF GROUNDED METAL)
LOWER INTERMEDIATE CLAMPING RING
BOTTOM PORCELAIN
BOTTOM CLAMPING RING
DRAIN PLUG
BOTTOM WASHER

a

b

Fig H-5 Typical porcelain bushings that may be replaced with polymers. **a.** Typical oil-filled bushing for 69 kV transformer. **b.** Sidewall mounted bushing

Lightning or surge arrester elements are enclosed in an insulated casing. Under severe operating conditions, or as a result of multiple operations, the pressure generated within the casing may rise to the point where pressure relief ratings are exceeded, The arrester then may fail, with or without external flashover, Figure H-6 exploding and violently expelling fragments of the casing as well as the internal components, causing possible injury to personnel and damage to surrounding structures. The action represents a race between pressures building up within the casing and an arcing or flashover outside the casing. The length of the casing of the arrester limits its ability to vent safely. The use of polymer insulation for the casing permits

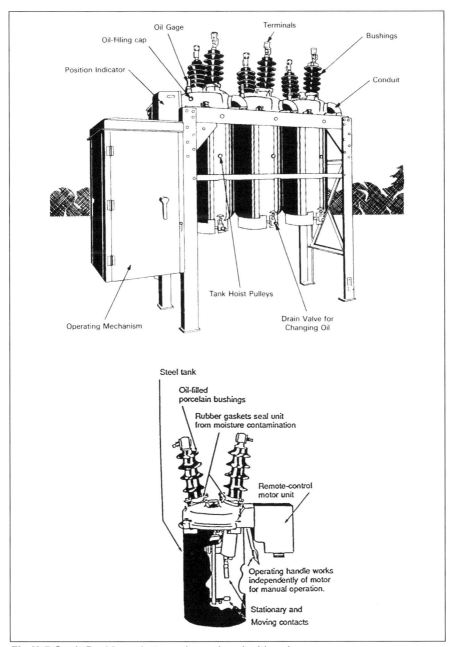

Oil Gage
Oil-filling cap
Terminals
Bushings
Position Indicator
Conduit

Tank Hoist Pulleys
Operating Mechanism
Drain Valve for Changing Oil

Steel tank
Oil-filled porcelain bushings
Rubber gaskets seal unit from moisture contamination
Remote-control motor unit
Operating handle works independently of motor for manual operation.
Stationary and Moving contacts

Fig H-5 Con't Bushings that may be replaced with polymers

245

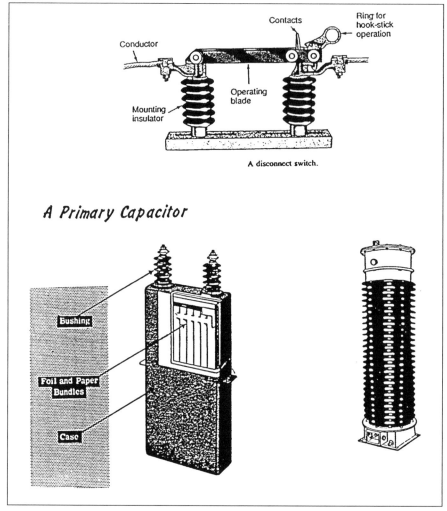

Fig H-5 Con't More porcelain bushings that may be replaced with polymers

puncturing to occur, without the fragmentation that may accompany breakdown and failure of porcelain

Summarizing, the advantages of polymers over porcelain include:

•   Polymer insulation offers benefits in shedding rain water or sleet, particularly in contaminated environments.

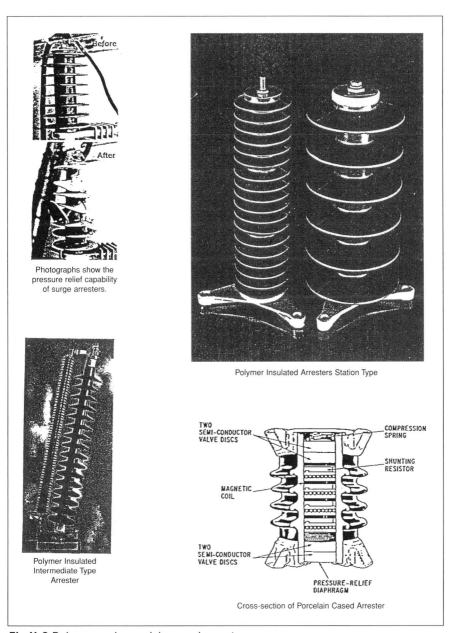

Photographs show the pressure relief capability of surge arresters.

Polymer Insulated Arresters Station Type

Polymer Insulated Intermediate Type Arrester

Cross-section of Porcelain Cased Arrester

Fig H-6 Polymer and porcelain cased arresters

247

- Polymer products weigh significantly less than their porcelain counterparts, particularly line insulators, resulting in cost savings in structures construction and installation costs. Table H-4.

- Polymer insulators and surge arresters are resistant to damage resulting from installation and to damage from vandalism. The lack of flying fragments when a polymer insulator is shot deprives the vandal from his satisfaction with a spectacular event and should discourage insulators as convenient targets.

- Polymer arresters allow for multiple operations (such as may result from station circuit reclosings), without violently failing.

- Polymer insulators permit increased conductor (and static wire) line tensions, resulting in lower construction designs by permitting longer spans, fewer towers or lower tower heights.

- Polymer one-piece insulators, lacking the flexibility of porcelain strings and the firm attachment of the conductor it supports are said to produce a tendency to dampen galloping lines.

Although polymer insulation has become increasingly utilized over the past several decades, there are literally millions of porcelain insulated installations in this country alone; economics does not permit their whole replacement. Advantage is taken of maintenance and reconstruction of such facilities to make the change to polymers.

Much of the data and illustrations are courtesy of Hubbell Power systems, and is herewith duly acknowledged with thanks.

Table H-4 Example - 10 Miles, 345 kV, 250 Strings of Insulators

Porcelain - 4,500 bells, 52-3, 13.5 lbs. ea., total 60,750 lbs. 750 crates at 3.1 cu. ft. = 2,325 cu. ft.
Polymer - 250 units, 14,4 lbs. ea., total 3,600 lbs. 5 crates at 75 cu. ft. = 375 cu. ft. Insulator cost = $51,750

		Savings
1. Storage space at receiving point (3 mos.) porcelain - 580 sq. ft.; polymer - 100 sq. ft.	480 sq. ft.	$60.00
2. Off-load, reload at receiving point porcelain - 10 man-hrs.; polymer - 2 man-hrs.	8 man-hrs.	$120.00
3. Breakage - off-loading, storage, reloading porcelain - 1 percent; polymer - 0	1 percent	$517.50
4. Truck - receiving point to tower sites (5 miles) porcelain - 1.00/cwt.; polymer .50/cwt.	$589.50	$589.50
5. Off-load at tower site porcelain - 5 man-hrs.; polymer - 1 man-hr.	4 man-hrs.	$60.00
6. Unpack at tower site porcelain - 50 crates/hour, 25 man-hrs.; polymer - 50 insulators/hr, 5 man-hrs	20 man-hrs	$300.00
7. Breakage - off-loading through string assembly & cleaning porcelain - 1 percent; polymer - 0	1 percent	$517.50
8. Assemble strings, attach blocks porcelain - 40 man-hrs.; polymer - 8 man-hrs	32 man-hrs.	$480.00
9. Clean insulators porcelain - 10 min./string; polymer - 3 min./string	29 man-hrs.	$435.00
10. Lift string into place (2 men) porcelain - 5 min./string; polymer - 2 min./string	25 man-hrs	$375.00
11. Install & connect to tower (2 men) porcelain - 5 min./string; polymer - 2 min./string	25 man-hrs	$375.00
12. Breakage - lifting & installation porcelain - 0.5 percent; polymer - 0	0.5 percent	$258.75
13. Cleanup packaging materials at jobsite porcelain - 6 man-hrs.; polymer - 1.5 man-hrs	4.5 man-hrs	67.50

Source: Hubbell Power Systems

Index

Ground current, 118
Ground neutral, 194-195
Ground relay, 118-119, 126
Ground wires, 55-56, 59-60
Grounding conductor, 213
Guy wires/guying, 11, 26-30, 68

H

Handling factors, 237
Hard-to-read meters, 234
Hazardous materials, 160
Heat detection, 46
Heat loss, 101, 181-182
Heavy-angle towers, 26, 68
Helicopters, 14
History, 139-140, 144, 146-147
Hollow cables, 10, 77-79, 87
Hollow conductors, 36-38
Hot spots, 46
Hot-stick methods, 65-66

I

Impedance, 189-190, 204:
　capacitance, 189-190
　inductance, 189-190
　resistance, 189-190
　resonance, 190
Incentive rate, 213
Independent power producers
　(IPP), 5-6, 124, 213
Independent system operators
　(ISO), 5, 213
Indicating instruments, 96
Induced voltage, 52
Inductance, 182-186, 189-190,
　204, 213:
　inductive reactance, 184-185

mutual inductance, 183-184
resistance and inductance,
　185-186
self-inductance, 183-184
Inductive reactance, 184-185
Infrared imaging, 46
Inspection/survey, 168
Instability, 122-123
Installation (cables), 80
Institute of Electrical and
　Electronics Engineers, 28
Instrument transformers, 106,
　108-109, 125, 213
Insulation value, 46, 237-238
Insulation, 6, 10, 15-17, 46, 76-77,
　79, 120-121, 237-249:
　fluids, 15
　materials, 15, 76-77
　value, 46, 237-238
Insulators, 8, 28, 33, 38, 46-68:
　assemblies, 47-50
　counterpoise, 57, 61
　fittings, 52
　lightning, 52-53, 56
　lightning/surge arresters, 53-55,
　　57-59
　maintenance, 64-67
　measurement methods, 59-63
　pin and post, 48-49, 51-55
　shield/ground wires, 55-56,
　　59-60
　surges, 63-65
　suspension insulators, 46-47
　tower footing resistance, 63
Integrated power delivery data,
　151-152
Integrated systems, 3